T0324931

Sterile Manufacturing

Sterile Manufacturing

Regulations, Processes, and Guidelines

Sam A. Hout

CRC Press
Taylor & Francis Group
Boca Raton London New York

CRC Press is an imprint of the
Taylor & Francis Group, an **informa** business

First edition published 2022
by CRC Press
6000 Broken Sound Parkway NW, Suite 300, Boca Raton, FL 33487-2742

and by CRC Press
2 Park Square, Milton Park, Abingdon, Oxon, OX14 4RN

© 2022 Sam A. Hout

CRC Press is an imprint of Taylor & Francis Group, LLC

ISBN: 9780367754358 (hbk)
ISBN: 9780367754372 (pbk)
ISBN: 9781003162506 (ebk)

Typeset in Times
by Deanta Global Publishing Services, Chennai, India

Contents

Preface

I contemplated writing this book on sterile manufacturing after spending over 20 years in the aseptic liquid and lyophilized parenteral drug vial fill/finish, cartridge filling for injector applications, sterile prefilled syringes, and ophthalmic bottle filling and capping. Quality by Design (QbD) dictates component preparations and compounding of drug formulations are key steps in cascading the bulk product from beginning to end. These processes all require extreme care and technical know-how in transferring across clean rooms. Personnel flow and materials flow separations and consideration for entering and leaving have rigorous requirements to ensure against cross-contamination and product safety. There are many guidelines that govern all these processes defining specific requirements. I wanted to put it all together in one accessible simplified reference to reinforce knowledge of regulations, processes, and provide an easy guide to follow. This book is important in providing a step-by-step understanding of what is required to manufacture sterile drug products and medical devices. The knowledge can be extended to the safe sterile production of food and clean cosmetics.

A brief description will show the reader how classified areas are functional in sterile manufacturing facilities. It will reflect on personnel behaviors, personal protective equipment (PPE), and process flow. Quality approach, the significance of systems validation to prevent cross-contamination and mixing is emphasized. Advances in disposable technologies and aseptic processing are highlighted. Critical utilities to generate water for injection and solution transfer systems are described as part of the integrated facilities design. Advances in controls automation to facilitate process reliability and faster production in a compliant manner are discussed. Reference to barrier and isolator technologies, and RABS to handle both API transfers and high potency drugs are key developments in this book. The systematic management of risk to ensure efficacious and safe pharmaceuticals is explained in terms of equipment qualification (EQ) and process validation. The importance of change control and analytical testing of raw material quality is systemic in drug manufacture.

For all intended reasons, the purpose of this book is to help my colleagues in the sterile production industries have a simple reference book on important issues that we deal with daily and always looking for ways to troubleshoot and solve problems that face us in manufacturing sterile products.

In writing this book I had to learn more and more about the subject matter that I am describing and covering, I had to investigate insights and real examples about situations that I had difficult and complex formulations that required specific innovative ways to process while ensuring compliance. I believe that through learning, training, and asking questions about alternatives, I was able to grow in how to convey this message in a more concise and focused fashion.

The feedback from the reviewers opened my thinking to how others are perceiving the materials and are required to add value in terms of case studies and real examples from experience.

Besides writing from my personal experience in the pharmaceutical industry, I had to research and follow guidelines from FDA, MHRA, EMA, ISPE, and ICH among other references that have been reviewed and cited as part of this publication.

I had many communications with colleagues to see how ideas can be explained and described in the desired methods. I have been thinking about how to aid on technical subjects that are pertinent to this field for over 6 years and I started putting together notes that would help my colleagues in the industry for over 10 years.

The table of contents in this book flows as intended for the reader to explore subject matter in a continuous fashion or refer to a specific section as needed. The main compliance message in this book is that FDA inspections focus – specific concentration on two areas: cGMP compliance and PAI for NDA. Documentation and procedures to cover WFI trending, pest control trending, overall facilities adherence to SOP and organizational structure, including appropriate protocols and reports covering all aspects of product submissions for pre-approval, namely exhibit batches data, stability batches data, microbiology and chemistry data – all presented in an orderly manner with full documentation including all CAPAs and deviations.

Acknowledgment

This book is written with the support of my family; my patient, calm, and loving wife Mona; my role-model daughter Samantha; and my inspirational son Owen. Thanks are extended to my brothers, my motivational cousin and lifetime friend Nael Hout, Ph.D., family, and friends at large in the United States, Lebanon, and other countries where they reside. This includes special acknowledgment for the intellectual and moral support of my brotherly friend Sarkis Khawaja. A special dedication is to the loving memory of my older brother, Dr Salim Hout, M.D. who influenced my upbringing and thought process by building unwavering support, self-confidence, and unconditional love. In addition, the transformational memory of my parents who supported our lives with everlasting compassion and support. I am very grateful to all my colleagues in the UK, US, and California who supported my growth and effort over the past three decades.

Author Bio

Dr Sam A. Hout, Ph.D., MBA, is an expert in pharmaceutical, medical, and health-care systems compliance. His experience spans over three decades in design, construction, and validation of cGMP manufacturing of drugs, medical devices, and delivery systems in clinical and commercial applications. As a senior consultant affiliated with Lachman, he provided cGMP inspection services both in the US domestic and international markets including PAI on new drug introductions. Previously, he was a senior director of engineering, project management, and process technology transfer at Siegfried, Irvine, CA. He held engineering and operations management positions at Teva pharmaceuticals, Phenomenex, Johnson & Johnson, and US government process chemistry research. In addition, his practice covered microbiology and chemistry laboratory operations including high potency drugs for oncology treatments with an emphasis on data integrity. For the past 20 years, his focus has been to design, validate, and operate sterile manufacturing facilities with automated aseptic and lyophilization capabilities.

1 Introduction

The purpose of this book is to highlight key ideas and factors to coach and guide professionals involved in learning more about this important sector in pharmaceutical and biopharmaceutical manufacturing facilities including operational requirements. This book will cover various regulations and guidelines instituted by the FDA, ISPE, EMA, MHRA, and ICH to emphasize GMP and PAI requirements in the sterile manufacture of medicinal products. This will also help compounding pharmacists and GMP inspectors and auditors. There are ample guidelines on this subject matter. The intent of this book is to present a simplified supplement with some very specific descriptions of aseptic processing techniques and methods based on personal experience and empirical knowledge.

Relevant and binding documents and information on sterile requirements could be researched in FDA aseptic processing guidelines, ISPE guides and baseline, ISO 14644-1/2, PIC/S, US Pharmacopeia, European Pharmacopeia, EU-GMP annex 1, and ICH guidelines. All GMP requirements apply to medicinal products for human and animal veterinary use.

Sterile manufacturing guides are intended to manage risk and introduce Quality by Design (QbD). The main intent is to minimize the risk of contamination due to mix-up, or improper cleaning by allowing microbes, particles, pyrogens, and endotoxins to be present in the process manufacturing steps.

Sterile manufacturing guidelines are based on specific elements in the construction and installation of classified clean rooms, critical utilities systems such as WFI, solution transfer systems, powder transfer systems, and monitoring and controls including automation of processes. Specialized technologies in fill/finish operations of vials, cartridges, bottles for ophthalmic or otic purposes, syringes (PFS), or ampoules, which might be filled and sealed aseptically or IV bags that are mostly terminally sterilized utilizing MES. In addition, considerations for personnel and material flow to prevent cross-contamination and are usually very well defined to separate in and out movements of sterile suites and cascading clean rooms through separate paths and including the use of path-through transfer systems for material flow.

Overall, paying attention to the design, construction, and installation should focus on process flow and employ standardization of equipment and methodologies as much as possible. Nonetheless, many batch manufacturing operating systems are customized to meet specific conditions of specialized products. Building sustainable premises, equipment qualifications, cleaning validation, and sanitization requirements should be all in consideration as part of the design of pharma/biopharma processing. Similarly, sterilization methods and periodic revalidation are all important factors that need to be on well-defined SOPs and schedules to ensure reliability of components used in aseptic filling and finishing.

The main emphasis of cleanroom classifications is to quantify particle limits when room operations are at rest (static conditions) and when in operation (dynamic conditions) for non-viable particles (see Tables 1.1 and 1.2). Microbiological limits are

TABLE 1.1
Clean Rooms GMP Classifications

FDA			COFEPRIS					EMA & WHO					ANVISA				
Maximum number of particles permitted/m³			Maximum number of particles permitted/m³					Maximum number of particles permitted/m³					Maximum number of particles permitted/m³				
In operation			At rest		In operation			At rest		In operation			At rest		In operation		
Class	.5 μm	5 μm	Class	.5μm	5μm	.5μm	5μm	GR	.5 μm	5 μm	.5 μm	5 μm	GR	.5μm	5μm	.5μm	5μm
ISO 5	3,520	NA	ISO 5	3,520	29	3,520	29	A	3,520	20	3,520	20	A	3,520	20	3,520	20
ISO 6	35,200	NA	ISO 6*	35,200	293	3,520,000	293	NA	NA	NA	NA	NA	NA	NA	NA	NA	NA
ISO 7	352,000	NA	ISO 7	352,000	NA	NA	NA	B	3,520	29	352,000	29	B	3,520	29	352,000	29
ISO 8	3,520,000	NA	ISO 8	3,520,000	2,930	3,520,000	29,300	C	352,000	2,900	3,520,000	2,900	C	352,000	2,900	3,520,000	2,900
NA	NA	NA	NA	3,520,000	29,300	Not defined	Not defined	D	3,520,000	29,000	Not defined	Not defined	D	3,520,000	29,000	Not defined	Not defined

TABLE 1.2
EN ISO 14644 Methodology

Class ISO 146144-1 (Federal Standard 209E)	Average Airflow Velocitym/s (ft/min)	Air Changes Per Hour	Ceiling Coverage
ISO 8 (Class 100,000)	0.005–0.041 (1–8)	5–48	5–15%
ISO 7 (Class 10,000)	0.051–0.076 (10-15)	60–90	15–20%
ISO 6 (Class 1,000)	0.127–0.203 (25–40)	150–240	25–40%
ISO 5 (Class 100)	0.203–0.406 (40–80)	240–480	35–70%
ISO 4 (Class 10)	0.254–0.457 (50–90)	300–540	50–90%
ISO 3 (Class 1)	0.305–0.457 (60–90)	360–540	60–100%
ISO 1–2	0.305–0.508 (60–100)	360–600	80–100%

TABLE 1.3
Microbiological Limits

Grade	Recommended Limits for Microbial Contamination			
	Air sample cfu/m³	Settle plate, Ø 90mm, cfu/4 hour	Contact plates, Ø 55mm, cfu/plate	Glove print, 5 fingers, cfu/glove
A	<1	<1	<1	<1
B	10	5	5	5
C	100	50	25	–
D	200	100	50	–

emphasized when classified rooms are in operation for viable particles (see table 1.3). Classification purposes are based on particle contamination with the understanding that classification is not monitoring. Media fills are conducted to certify rooms for aseptic conditions to confirm the normal production of drug products.

In writing about sterile and aseptic manufacturing, all information is presented as guidance to assist in pharmaceutical manufacturing. Therefore, the author cannot ensure, guarantee, or warrant that using the information in this book will result in full acceptance by regulatory agencies, as there are many factors that are in play for registration.

From the aseptic core to the controlled-non-classified (CNC) sanitary surrounding corridors, there are minor variations among the standards that are pronounced in table 1.1. Therefore, the manufacturer of a specific product should comply with the requirements of the country where their standard applies for commercialization of a drug product in that country.

I will attempt during the course of the following chapters to exhibit and discuss – Case Studies/potential challenges of complex formulations – I will attempt to add some specific practical application/case study that I can choose from my experience in aseptic manufacturing. I will weave these examples in some of the related subject chapters. This will cover some challenges that were overcome as it relates to complex formulations and API preparations.

Advances in restricted access barrier systems (RABS)/depyrogenation tunnel – Barriers & Isolators (RABS) will also be covered. FDA inspections focus – specific concentration is illustrated in two areas: cGMP compliance and PAI for NDA. Documentation and procedures to cover WFI trending, pest control trending, overall facilities adherence to SOP, and organizational structure to cover:

- C&Q of critical systems and equipment
- Process validation
- Room cleaning
- EMPQ
- Drawings
- Calibration
- Equipment PMs
- Supply procurement
- SOPs/ WI/OJT
- Training plan/rosters
- Training updates
- Documents updates
- Batch records/electronic batch records
- CSV
- QMS/protocols/reports

2 Data Integrity Compliance

procedures for compliance with 21 CFR Part 11 – must adhere to Data Integrity compliance, e.g. all lab instruments, process control systems (PLC or HMI), computerized systems, and excel spreadsheets used should all comply with data collection, storage, and usage requirements such as:

1. Programmable Logic Controllers (PLC) or other intelligent processors
2. Human machine interface (HMI)
3. Analytical instrument software
4. Computer systems

Computer System – A combination of computer software (including quality excel spreadsheets), hardware, peripherals designed to perform a specific function. Examples of computer systems include process control (equipment, environment); business (accounting, enterprise resource management); and database (control charting, LIMS).

Critical Data – Critical data (including parameters) is defined to be any data created, modified, or maintained by a system/instrument/PLC/HMI which will support quality decisions related to product safety, efficacy, and the quality of the product such as:
- Non-conformance Investigation
- Modifications/maintenance activities for equipment
- Product release requirements
- PLC or HMI parameters that could potentially impact product acceptability.

Data Integrity – the extent to which all data are complete, consistent, and accurate throughout the data lifecycle. Complete, consistent, and accurate data should be: Attributable, legible and permanent, contemporaneously recorded, original or a true copy, and accurate (ALCOA).

Dynamic records – A dynamic record is a record format that allows interaction between the user and the record content. Examples of a dynamic record include chromatographic data file, which allows processing to change the baseline, or a spectral data file, which allows changes in the displayed output.

GXP – General term used to represent multiple regulations defined by Regulatory Agencies (e.g. Good Manufacturing Practices (GMP), Good Clinical Practices (GCP), Good Laboratory Practices (GLP), etc.) that are applicable.

Information Management – the group responsible for the management of a particular computer system.

Metadata – Metadata is the contextual information required to understand the Original or Processed Data. Metadata is often described as data about the data. Metadata is an integral part of the original (raw) data or process data (without metadata, data has no meaning) Metadata for a particular piece of data could include the date/time stamp for when the data was acquired, a user ID of the person who generated the data, the instrument ID used to acquire the data, etc. An audit trail is a form of metadata.

Predicate Rule – Federal regulation that defines the requirements for record creation, retention, and/or signature/individual identification.

Static records – A static record is a fixed-data document such as a paper record or an electronic image. Examples of a static record include a paper printout from a balance, or a static image created during the data acquisition.

Local Role	Responsibilities
• Software Quality Engineer	Responsible for the overall management of the
• Quality Lab Associate	compliance program for areas assigned, maintenance of inventory list, leading coverage and gap analysis efforts, and remediation activities (if applicable).
• Area Manager	Responsible for approvals of coverage
• Quality Engineer Manager/Quality	assessments for 21 CFR Part 11 as well as
System Representative	Data Integrity assessments upon completion.
• Computer System Owner/ Business	Responsible for assisting with coverage
Process Owner	assessments for 21 CFR Part 11 and Data integrity.
• Information Management	Responsible for assisting in inventory and analysis efforts and administration of remediation process as required.

21 CFR Part 11 – Electronic Records

21 CFR Part 11 defines the requirements for the use of electronic records and electronic signatures which are used in lieu of paper records (21CFR Section 11.2) in which the records are required per GXP procedures. Any laboratory instrument, PLC, or computerized system which retains electronic records that are used to replace existing paper records is considered to be covered by the requirements of 21 CFR Part 11 for electronic records. Systems that maintain electronic records and the intended use is paper are not considered to be covered by 21 CFR Part 11, although the system must comply with all data integrity requirements.

21 CFR Part 11 – Electronic Signatures

21 CFR Part 11 defines the requirements for the use of electronic signatures associated with electronic records if the signature is used to replace a signature recorded on paper. Any laboratory instrument, PLC, or computerized system that is used to replace paper-based signatures are considered electronic record systems and must comply with all requirements of 21 CFR Part 11 for electronic signatures. Electronic signature training must be documented and verified prior to adding new users to an electronic signature system.

Data Integrity

Any laboratory instrument, PLC, or computerized system that retains GXP relevant data (regardless of Part 11 Coverage) must comply with all data integrity requirements to ensure data is complete, consistent, and accurate throughout the data lifecycle.

PROCESS/PROCEDURE – INVENTORY OF SYSTEMS

An inventory of all laboratory instruments, PLC, or computerized systems will be maintained for all systems (laboratory, business, and process control) in use. The following information is required for each system:

a. System name
b. Functional description
c. Platform (MES, Lab Instrument, PLC, HMI)
d. Location
e. Business process owner
f. 21 CFR Part 11 coverage
g. GAMP category (category 1, 3, 4, or 5)
h. Records maintained required by GXP (Yes or No)
i. Types of GXP relevant data maintained by the system
j. Validation specification reference
k. Risk classification
l. System classification (category)
m. System risk classification (high, medium, low)
n. 21 CFR data integrity compliant
o. Gaps identified (if any)

PROCESS/PROCEDURE – DETERMINATION OF 21 CFR PART 11 / DATA INTEGRITY APPLICABILITY

PROCESS/PROCEDURE – 21 CFR COVERAGE ASSESSMENT

All laboratory instruments, PLC, or computerized systems must be evaluated and classified for quality system impact, 21 CFR Part 11, and data integrity applicability. The results will be recorded in the computer system inventory.

Process/Procedure – 21 CFR Part 11 / Data Integrity Gap Analysis

Any laboratory instrument, PLC, or computerized system that is covered by 21 CFR Part 11 must complete a gap analysis to determine if the system is fully compliant with all requirements. Any gap identified must be addressed procedurally or with a system change to bring the system into full compliance prior to implementation.

Change History

Change #	Description of Change
	Add excel spreadsheet clarifications and new sections for coverage assessment and gap analysis.
	Combine forms X and Y into Z form. Also, update inventory elements as required.
	All sections updated for data integrity requirements. For example: Formatting change only. Obsolete form X. Create new form for data integrity evaluation Y. Re-process Form Z, Add definition of critical data – Alphabetize definitions. Form Z – Added additional questions for assessment related to Autosave and security.

3 Risk-Based Life Cycle Management

The Risk-Based Life Cycle Management (RBLCM) process is a risk-based evaluation of pharmaceutical drugs, Medical Products (MP), devices, and therapeutic products that ensures adherence to product and regulatory requirements. The process assesses current manufacturing, design, and documentation. In addition, product and process performance data is collected to determine if the current design, manufacturing processes, and risk controls continuously mitigate failure modes that could lead to unacceptable harm to a patient/user or could impact a therapy.

Terminology	Definition
Control Strategy	A planned set of controls, derived from the current process and product understanding that assures process performance and product quality.
Critical Control Point	Point in manufacturing at which controls are applied and data is generated to prevent, eliminate, or reduce the risk related to Essential Requirements (ERs) to an acceptable level.
Essential Requirements	Design requirements (characteristics or attributes) of the product that, if not met, can result in harm to the end user. These are provided by the product design owner.
End User	Any internal or external (e.g. patients, caregivers, bystanders, service technicians, environment) user of a product or a process used to manufacture a product.
Hazardous Situation and Harm Analysis (HSHA)	The HSHA is a systematic tool for the analysis of hazards and hazardous situations specific to a particular therapy to identify the resulting harm to the end user (e.g. patient/caregiver) and the probability of severity of harm occurring. This assessment is predicated on a particular therapy and a variety of patient populations. It is also predicated on the types of products used in that particular therapy.
Listening Systems	Internal or external data systems that provide feedback on either product or process performance.
Life Cycle	All phases in the life of a product from the initial conception to final decommissioning and disposal.
Other Requirements	Non-essential, non-regulatory design requirement that potentially impacts product quality and/or production.
Risk-Based Life Cycle Management (RBLCM)	A risk-based evaluation of Medical Products (MP), devices, and therapeutic products that ensures adherence to product and regulatory requirements.

(*Continued*)

9

Terminology	Definition
Playbook	A collection of documents including manufacturing risk analysis, Process Failure Mode and Effects Analysis (PFMEA), Control Plan, Product Performance Data, and process and test method validation assessments, created for a finished goods product family or a product manufactured at a specific site, giving documented evidence that the product and process are in the state of control. The playbook deliverables are also used to identify the need for mitigation or improvement activities.
	Also, a collection of risk-mitigating living documents that will be continually improved within the context of the Quality Management System as product knowledge evolves.
PDO (Product Design Owner)	Personnel with the responsibility for issuing Product Requirements for a specific product family.
Progression Level	Defined content and scope of the playbook. Higher levels are related to increasing product knowledge.
Product Characteristics	Product characteristics are the measurable attributes of a process step or sub-step. Product characteristics can be found in engineering specifications, drawings, etc. Examples include Dimensions, size, and tensile strength.
Product Family	A grouping of products that have the same or similar intended function/ indicated use, fundamental technology, performance specifications, and/or use in practice and, in general, with exceptions, only differ in non-essential characteristics that do not affect safety and effectiveness.
QSIP (Quality System Improvement Plan)	A global improvement effort intended to create a new culture that drives continuous improvement through proactive control and mitigation of risk. QSIP is supported by three main pillars: RBLCM, Global Product Ownership, and Quality Quotient.
Regulatory Requirement	Non-essential design requirements related to a regulatory obligation (e.g. commitment, guidance, standards). If a regulatory requirement is not met, the product may be considered adulterated.
Severity	A measure of the possible consequence of a hazard (the consequence of a hazard is the harm to the end user).
State of Control	A condition in which the set of controls consistently provides assurance of continued process performance and product quality. Product use test results, complaint data scrap, and internal process data are examples.
Therapy	The treatment of physical or mental illness.
Process Failure Mode and Effects Analysis	The PFMEA considers all reasonably foreseeable potential failure modes of each manufacturing process, their causes, and effects on the product.
Linkage Document	A tool utilized to relate process steps and their associated process parameters to requirements. This tool aids in the creation of the PFMEA.
Process Flow Diagram	A visual summary of the process steps required to manufacture a product.
Control Plan	A plan documenting the manufacturing process controls for product and process characteristics.

The playbooks are to be created by cross-functional teams including representatives from plant quality and plant manufacturing, as necessary, including process and product subject matter experts (SMEs). The purpose of the cross-functional team(s) is to conduct a comprehensive analysis of each product family. Under no circumstances shall these deliverables be created by individuals.

Member	Responsibilities/Authority
RBLCM Core Team	Establishing the RBLCM process/procedure
	Approving changes to the RBLCM process/procedure
	Maintaining alignment of tools and content across product families and manufacturing sites
Product Design Owner (PDO)	Participating in the development, update, and approval of essentials
	Participate in the development of the product characteristics
	Participate in reviews of RBLCM outputs (e.g. control strategy assessment, gap summary) as required
	Requirements responsible for establishing the Essential Requirements
Plant Team Lead Equivalent	Selecting the RBLCM plant team
	Defining the scope of the analysis
	Ensuring that technical content is provided by qualified team members
	Coordinating the cross-functional team conducting a risk assessment
	Reviewing change documents for RBLCM effect
Quality/Manufacturing/ Engineering Representatives	Participating in the development and approval of RBLCM deliverables
	Participate in the development and maintenance of FMEA's Control Plans, etc.
Contributing Functions	Providing information according to their area of expertise
Business Quality Leader (Plant Manager and Plant Quality Director)	Approve and communicate the final decision of Gap Summary and Disposition
Franchise Specific Disposition Team	Assess gaps to determine major/minor categorization and provide disposition recommendation

The goal of RBLCM is to:

- Establish objective evidence that demonstrates products and processes are operating in a "state of control". The objective evidence created shall be documented in a product family "playbook" for a single product family or product manufactured or serviced at a unique Medical Products site
- Periodically analyze the documented risk controls (control strategy) for effectiveness and consistency within and across product families
- Maintain selected "playbook" deliverables as living documents by continuously enhancing the content as new product/process information is obtained
- The playbooks serve as mechanisms for knowledge sharing and improvements globally
- Manage and reduce risk, increasing the overall health of product families by driving actions based on ongoing listening system data monitoring and trending, e.g. defect per million (DPM) and complaints per million (CIPM)

The RBLCM process consists of the following main elements:

- Creating playbook documentation using RBLCM methodology

- Monitoring and trending product performance for each Essential Requirement over time
- Analyzing playbooks for adequate risk control within the product family and across plants
- Updating playbook documents related to Essential Requirements as new information is obtained

The RBLCM playbook is a summary of objective evidence within the manufacturing space that not only demonstrates that legacy product's design and manufacturing processes meet the requirements of the customers but also demonstrates that products are compliant and safe. This is based on the characterization of the manufacturing processes and the in-process, DPM, post-production, CIPM, and listening systems. The RBLCM playbooks are created for each finished goods product family manufactured at a specific site.

The "playbook" and other related documents consist of the following elements:

- Product Overview (based on product family)
- Essential Requirements list
- Process Flow Diagram
- Process Linkage Document (historical document)
- PFMEA
- Control Plan
- Assessment of existing Process Flow Diagram, PFMEA, Control Plan
- Assessment of Test Method Validation (TMV)
- Product Performance Analysis
- Assessment of Process Validations
- Quality Plan
- Gap Summary and Disposition

All playbooks generated in the site facility will follow the requirements outlined in this document.

- Identify product families
- Identify requirements ERs for each product family
- Identify key plant processes that impact the identified requirements
- Complete Trace Matrix, PFMEA, and Control Plan for each key plan process impacting identified requirements
- Define internal and external product performance for the identified requirements
- Review validations and test methods for the adequacy
- Identify gaps, document the Quality Plan, act to address, ensure containment if required

The RBLCM process scope is defined by progression levels (PLs):

- Level 1 – Level 1 is the state before implementing the RBLCM process
- Level 2 – Regulatory compliance state: Creation of initial documentation (objective evidence) to demonstrate the state of control for processes and products. Limited use within the life cycle. Level 2 is having completed RBLCM for initially defined Essential Requirements

- Level 3 – Product performance improvement: Move from a reactive to a proactive state. Expanding documentation, scope, life cycle integration, and process analysis to proactively determine product or process issues and improvement opportunities
- Level 4 – Continuous process monitoring: Continue to progress toward a preventive state: Further expanded scope of manufacturing stream and documentation to create a robust variation reduction management system
- Level 5 – Business ecosystem: Proactive, risk-based system, with holistic integration of quality, manufacturing, commercial, and R&D systems

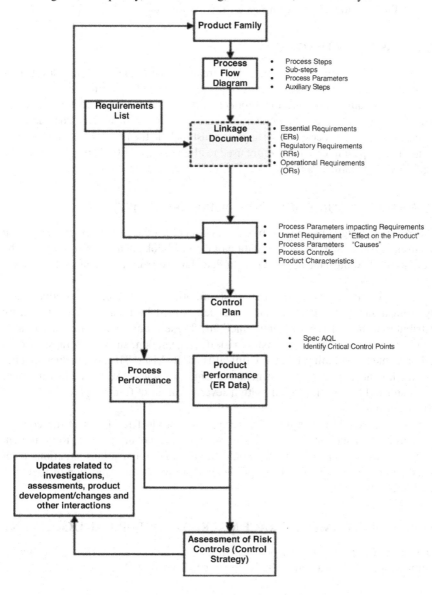

The product family overview document includes, at a minimum:

- All applicable product codes manufactured at the site
- A description of the product (include pictures, illustrations, etc.)
- The relevant product history
- A description of how the product is used (intended use)
- Where the product is sold or distributed
- The manufacturing volumes (# units annually) or number of units in the field
- Other manufacturing locations for this product family

PROCESS FLOW DIAGRAM

The Process Flow Diagram is a visual summary of the sequence of high-level steps or processes, beginning with receiving and inspection and continuing to the production release required for a salable product. Only operations that transform material into subsequent intermediate products to achieve the finished good are considered as process steps. The Process Flow Diagram must include the process step number and name. The step number and name are used to align other RBLCM output documents, e.g. PFMEA, Control Plan.

ESSENTIAL REQUIREMENTS (REQUIREMENTS LIST)

The Product Design Owner (PDO) provides a set of Product Requirements (essential requirements) that will be the foundation of the playbook documents. These describe the Essential Requirements by each product family from a patient and caregiver safety perspective.

Essential Requirements are the most important aspects of the product family from a safety perspective. They are determined by relating Critical Quality Attributes to design requirements/functions. These requirements are then related to hazardous situations from existing risk documentation, such as therapy-specific risk documents including HSHA, Clinical Hazards List (CHL), or product-specific risk documents (e.g. Risk Assessment and Control Table (RACT)), that, if not met, can result in harm to the patient with a severity rating of Catastrophic (5), Critical (4), or Serious (3).

Essential Requirements must be defined at a product level (not at the component, process, or subsystem level). The non-achievement of an Essential Requirement is an end effect on a product and should be relatable to hazardous situations from risk documentation. The PDO will guide and approve the identification of Essential Requirements.

LINKAGE DOCUMENT (TRACE MATRIX) – HISTORICAL DOCUMENT

The Trace Matrix/Linkage Document links the ERs to key steps of the manufacturing process/parameters. This document lists the ERs across the top of the document and process steps and parameters along the side. This document is used to highlight the process parameter relationship and a high-level overview of the process step.

- ERs across top
- Process steps along the side
  - Parameters related to process step function/failure
  - An "x" will identify the process step to the Essential Requirement potentially affected
- Every column (requirement) linked to a process step/parameter will be identified with an "x" at the intersection of the column (ER) and row (process parameter/step)
- The linkage document is used to highlight the relationship of the ER to the process. This will assist with the initial PFMEA generation

The Linkage Document is a tool utilized to relate process steps and their associated process parameters to requirements. This tool aids in the creation of the PFMEA. The matrix documents the relationship between the process steps and their process parameters that may affect an Essential Requirement. This relationship is determined and noted within the matrix.

Process steps should be aligned between the Linkage Document and PFMEA.

PROCESS FAILURE MODE AND EFFECTS ANALYSIS

The PFMEA considers all reasonably foreseeable potential failure modes of each manufacturing process, their causes, and effects on products that can affect the outcome of an Essential Requirement. Based on the failure mode and the causes, the related manufacturing controls, both prevention and detection, are documented. At a minimum, the PFMEA must be completed for all process steps and process parameters listed within the Linkage Document that impacts ERs.

The PFMEA identifies the risk associated with the manufacturing steps that can cause the Product Requirements to be non-conforming. It also identifies the controls that are currently in place to mitigate those risks. The FMEA is to start at the beginning of the process and continue in sequence until the process end. It will identify the process step, Risk (Failure mode), Risk Identification (Controls), Risk Analysis, and relationship to the ER.

- Severity numbers come from the list of Essential Requirements, given by PDOs. In order to emphasize the overall review process, the severity number for Essential Requirements has been set to the highest rating of 5.
- Detection rating is assigned based on control methods currently in place. The occurrence rating is based on the estimated defect level that is presently considering the preventative measures that are in place.
- The PFMEA is used to identify the risk in the process and also identify the key controls for the potential failure mode/defect. A risk of 1 is assigned to

the optimal condition, while a 5 is assigned to items considered to have the highest risk, occurrence, or lowest ability to detect. Optimal RPM $1 \times 1 \times 1 = 1$, worst RPN $5 \times 5 \times 5 = 125$.

The overall rating (RPN) is the Severity $\times$ Occurrence $\times$ Detection

CONTROL PLAN

The Process Control Plan (Control Plan) details the product and process controls from the PFMEA.

The Control Plan is to document elements of the control strategy utilized at each step to meet the Essential Requirements and minimize process and product variation. It is complementary to the PFMEA and details the prevention and detection risk reduction measures (controls) identified in the PFMEA. At a minimum, the Control Plan must include and be completed for all process steps and process parameters listed within the pFMEA. The Control Plan must include the following information:

- Process step numbering and process name/description consistent with the pFMEA
- Product and process characteristics related to Essential Requirements per process step
- All in-house processing, inspection, packaging, and release steps that can affect the outcome of an Essential Requirement
- Equipment for process and testing
- Reference documents for product characteristics
- Validation references for process parameters and test methods
- Characteristics classification for products and processes
- Product specifications and process tolerances that will be maintained in quality documents, summarizing information for the product family
- Inspections
- Sample size, frequency, and person(s) responsible for testing or inspection
- Control method
- Reaction Procedure

ASSESSMENT OF PROCESS FLOW DIAGRAM/ PFMEA/CONTROL PLAN DOCUMENTS

The assessment is used to evaluate existing Process Flow/pFMEAs and Control Plans against an established list of minimum requirements to ensure compliance with the RBLCM process. If gaps are identified within the initial assessment checklist, corrections must be made to the Process Flow Diagram/ pFMEA/Control Plan documents. Upon making the corrections (or creating new document(s)), the assessment will be completed by indicating that all requirements have been met.

ASSESSMENT OF TEST METHOD VALIDATION DOCUMENTS

The Test Method Validation Assessment evaluates existing Test Methods for each Essential Requirement against an established list of minimum requirements to ensure compliance. If gaps to the minimum requirements are identified while reviewing a test method, they must be documented as a gap in the Quality Plan.

ASSESSMENT OF PROCESS VALIDATION AND MANUFACTURING INSTRUCTION DOCUMENTS

The Process Validation and Manufacturing Instructions Assessment was conducted to determine if the processes used in the manufacturing of the product family are correctly validated. This assessment confirms the quality of the validations against a standard provided by the core team. Gaps not corrected will be recorded in the Quality Plan.

The Process Validation and Manufacturing Instruction Assessment evaluates existing Process Validations for each process step identified in the Linkage Document that impacts an Essential Requirement. These process steps are evaluated against an established list of minimum requirements to ensure compliance. If gaps to the minimum requirements are identified while reviewing a process step, then the gap and mitigation must be noted within the assessment sheet. All gaps not meeting the assessment's minimum requirements must be added to the Quality Plan. This does not include benchmark items.

PRODUCT PERFORMANCE DATA

The Product Performance Data details the product performance of the ERs and other important requirements with internal and external measures.

- Data will be reviewed from Final Inspection Test Data and Complaints
- The 95% confidence bound will be calculated from the total defects and total samples
- These upper bound limits will be compared to a 0.065% level and the current AQL defined in the appropriate product specification
- Any item that does not meet the 95% confidence level of the product AQL will be identified as a gap in the Quality Plan and highlighted

This data will be updated quarterly. The Product Performance Data is the objective evidence that demonstrates a product family is safe and compliant with current acceptance criteria. Each Essential Requirement shall have its product performance calculated. The data will be trended after progression level 2 completion. The focus is on the collection of data (variable or attribute) from the final product (finished goods) testing. If no final product testing is conducted, upstream testing where the requirement was last tested can be utilized. If neither finished good or in-process testing exists, receiving and inspection testing may be used.

The Product Performance Data report includes:

- Primary Method: Product data generated from the final product, in-process is compiled over a time period to make a statistical inference regarding how the Essential Requirements met the specification AQL. Sufficient data needs to be collected to allow a statistical inference to be made, that is, enough data to obtain a 95% Confidence Limit that the DPM is less than the Specification AQL. The defect rate for each Essential Requirement is calculated for attribute data by using the number of defects over the number of samples and reported as DPM (upper bound 95% defective). Process capability is calculated for variable data and reported as PpK or based on variable capability data and converted to DPM
- Any failure to meet the current acceptance criteria will be documented as a gap in the Quality Plan. If an existing CAPA exists for a known problem, the CAPA number will be referenced in the appropriate section
- Customer complaint data aligned to Essential Requirements are used to calculate the complaint rate for each Essential Requirement as the number of complaints instances over released units
- Create a chart or multiple charts that will show the product performance (% defective/DPM or PpK) over time (for each measurement interval)

PRODUCTION PROCESS DATA COLLECTION (PPDC)

PPDC can refer to any production data, including Requirement, Product Characteristic, and Process Parameter Data. As part of Progression Level 3, the Production Process Data Collection (PPDC) identified shall include the data monitoring of Critical Control Points (CCPs).

CCPs may include Essential Requirements, Product Characteristics, or Process Parameters. Analyzing data for CCPs as part of PPDC at Progression Level 3 is the primary mechanism to describe Process Stability for the State of Control.

PPDC requirements for PL3 (in progress):

- CCPs shall be referenced in the Control Plan
- Data shall be managed per local data management processes
- Any CCP data collected shall have appropriate control/alert limits defined
- Control/alert limits shall be within specification limits and derived from process variation as opposed to process capability
- Manufacturing data sources for CCPs must have objective evidence for Test Method Validation (TMV)
- Data shall be collected for CCPs at least weekly
- Rationale for selection of CCPs shall be included in the control strategy assessment

The Quality Plan represents a summary of any gaps and the immediate containment action(s) (i.e. additional or different inspection techniques, increased sample size, restricted ranges of operating values, etc.) that will remain in place until the gaps are addressed. The Quality Plan captures gaps and their mitigation plans identified during the execution of the process.

- CAPA Activity associated with the Essential Requirements gaps identified
- Defining/addressing gaps
- Trackwise (computerized change control system) nonconformance

The Gap Summary and Disposition Document details a summary of the gap(s) identified within the Quality Plan during the RBLCM process for a particular Product Family. The gap(s) potential impact on the customer is identified and ranked as either major or minor.

- Major Gap: A gap in a product, process, and/or quality system that could potentially result in the manufacture and release of a device or therapeutic that does not meet one or more of its requirements/specifications, and the product defect would have a severity of Serious, Critical, or Catastrophic
- Minor Gap: A gap in a product, process, and/or quality system that could potentially result in the manufacture and release of a product/device or therapeutic drug being that does not meet one or more of its requirements/specifications, and the product defect would have a severity of minor or negligible. Furthermore, this can include a documentation gap that would not affect the ability of a released product to meet its requirements/specifications

Gap Summary and Disposition Rating

Rating	Disposition Description
1	No gaps have been identified
	No product impact
2	Only Minor gaps identified. No product safety impact
3	Major gaps identified; however, benefits outweigh the risk of the gaps identified (rationale required)
4	Major gaps identified; full risk–benefit analysis required to continue releasing the product

Disposition 1 or 2 does not require a written Rationale and/or Action to support the Overall Assessment and Dispositions. In contrast, Disposition 3 or 4 requires a written Rationale and/or Action to support the Overall Assessment and Dispositions. Rationales for the determination of acceptable or unacceptable risk should be based on the risk management process. Disposition 4 requires a risk–benefit analysis. Based on the information analysis provided and additional pertinent information, a decision is made to determine if the containment and mitigation activities in place provide an acceptable level of risk to the patient based on the benefit of therapy provided.

The Gap Summary and Disposition document shall be reviewed with Franchise Specific Dispositioning Team member(s). The team will review and document the potential impact on the patient and categorize the gap(s) identified for a Product Family as Major or Minor.

The Regional Business Quality Leader and cross-functional team member(s) document the rationale for major gaps with a disposition of 3 or 4, if applicable, and sign the Gap Summary and Disposition document.

ASSESSMENT OF RISK CONTROLS (CONTROL STRATEGY)

The RBLCM process documentation requires a periodic assessment. At a minimum, this assessment shall be completed once the other requirements have been completed and annually thereafter. The assessment shall evaluate the failure modes/causes and their related process controls (prevention and detection), along with the product performance and process data of the requirements, product characteristics, and process parameters to evaluate effectiveness. This analysis will provide a foundation to compare controls across product families, plants, equipment, suppliers, service centers, etc.

MAINTAINING "LIVING DOCUMENTS" UPDATES
TO THE PLAYBOOK DOCUMENTATION

The playbook contains information that documents product requirements, product or process failures, causes of process failures, preventive and detection controls, and product performance, among other information associated with a particular product family and its manufacturing process. Therefore, this knowledge can be used for:

- Identify improvement opportunities
- Data/Information Monitoring – to review data, information, and associated trends to determine if an event or change has occurred to product and/or process
- Events – to identify, investigate, and evaluate events
- Changes – to evaluate, implement, and monitor the effectiveness of changes including new products/processes

The playbook deliverables are to be considered "living" documents, and as such they shall be updated as new information related to the requirements is obtained. Potential sources or triggers that may require a change in the content of the "playbook" deliverables are summarized.

Potential Sources of Triggers to Update Playbook

General Description of Trigger	Examples	Requirements
Product Family	ID New Gap	Event
GAP	Gap Closure	
New Product/Requirement or Change to Requirements	Code Merged/Added into Product Family	Changes
		Improvements
	Product Line Extension	
	Regulatory Requirement	
	Performance Requirement	
	Essential Requirement	
Proactive Period Review	Annual Product Review (APR)	Monitoring
	Risk Review	
	Management Review	
	Plant Reviews	
Product Family Code, Volume, Description Change at Plant	Change in codes manufactured at plant associated with exiting product family	Changes
	Change in volume	
	Update to the product description	
Nonconformance in Equipment, Calibration, Raw Material, or Process	AQL Failure	Events
	OOL/OOT	
	Supplier Defect	
	Audit Observation (Internal/External)	
	Equipment Set-Up	
Nonconformance of Released Product	Complaint	Events
	Adverse Event (AERs)	
	Field Alert Report (FAR)	
	FCA/Holds	
	Distribution Conditions	
New/Improved: Product, Process, Facility, or Supplier change	New Supplier	Changes
	Supplier Notice of Change	Improvements
	Equipment Change	
	Material Change	
	Label Change	
	Validation/Qualification	

DOCUMENT MANAGEMENT

The initial RBLCM playbook is approved and filed as a protocol document. As part of the progression of the RBLCM process, a Quality Management SOP system (RBLCM) is being used for updating the playbook sections.

The RBLCM playbook will be a living document that will be maintained by the specification department and is viewable online. The product overview, ERs, Trace Matrix, pFMEA, Control Plan, and Process Performance Data will be listed.

The documents may be updated based on updated controls, process changes, validation activities, CAPA and complaint investigations, etc.

Updates to the Performance Section are performed by quarter. This update to performance is to provide a larger amount of data for trending. This requires an annual review/audit for current references and compliance. Playbook sections are approved per the RBLCM.

PLAYBOOK UPDATES/APPROVALS (LIVE SYSTEM)

The Process Flow Diagram, pFMEA, Control Plan, and Process Performance Updates must be reviewed as required. Changes to the product overview, test method validation, and process validation assessment will be approved by Quality engineering, RBLCM Representative, and other additional reviewers as needed. Changes to Essential Requirements will require approval from the product design owner, plant manager and Quality manager, and the RBLCM Representative.

The playbook will be controlled, which includes storage and change management. The SME (RBLCM Representative) and approver will verify the affected areas. The initial playbooks are stored as protocols and are maintained in the Site Doc Center.

RBLCM DATA COLLECTION

RBLCM data is collected for monthly trend reports as well as quarterly Product Performance Data trend reports. The following data is collected for trending:

- Product Use Testing
- Packing Defects

Product Use Testing data and leaker data from the Leaker Analysis system are also collected. Control records are created to link certain defects to the ERs and to group product codes by families. The exported data are linked to these control records for defects, and complaints can be trended by the product family for each Essential Requirement. Batch production totals are obtained to calculate DPM and complaints per million. File security is applied to the new data to ensure it is not modified once it is verified.

4 FMEA Manufacturing Procedure

This outlines the relationship between the individual process (equipment) failure mode and effects analysis (FMEA) and the products produced. Each manufacturing area has a table that aligns the process FMEAs required for each product type. This is completed for all manufacturing areas: component preparations, filling, sterilization, and packing. The combination of various area FMEAs provides the overall risk assessments by product type. This procedure is used in combination with several other site procedures to understand and perform risk assessments. In order to assess manufacturing risks for site manufacturing/production equipment processes, FMEAs are created and documented. These FMEAs for components, filling, and packing production areas are generated as per requirements. In addition, the overall risk assessment is also completed per procedure. This risk assessment is a start to the end of facility assessment that focuses only on essential requirements. The systems complement each other by allowing independent reviews to be completed.

An overall High Level Risk Analysis (HLRA) govern an overview of Site systems, processes and tools that comprise the risk management system.

Other requirements related to the Site facility validations outlines procedure FMEAs related to manufacturing equipment and processes. Sterilizer vessel qualification and validation process parameters are covered under specifications, and sterilization process FMEAs

Site FMEAs are assigned unique control numbers. If an existing FMEA is updated, the previous issue becomes superseded. Change history is documented. Site manufacturing-related FMEA are categorized by product types. These categories are based primarily on product size and represent the different product types manufactured at Site. As FMEAs are created or revised, requires Quality Engineering (or the FMEA Initiator) to determine if FMEA needs to be updated. Potential reasons for updates include: new equipment, new technology, product/process change, equipment modifications, etc.

All classified environments in which intermediate and final products are manufactured or manipulated, including the storage of components and solutions and the product itself, have been evaluated per Environmental FMEAs. Environmental FMEAs are used as part of the overall environmental monitoring program. The Environmental FMEAs are controlled per this procedure.

- SITE – Environmental FMEAs are the risk matrix for ranking the three risk categories of "Likelihood of Transfer", "Source Condition", and Detection/Controls". A physical review of the applicable area is required when updating the FMEAs in order to observe the source condition of the process, room, etc. This rating scale is sensitive to the evaluation of an area for environmental assessments in comparison with the process rating scale where the occurrence is measured, severity is based on AQL, and occurrence on a physical test. Environmental FMEAs are risk assessments maintained as living documents and must be reviewed when a change is performed per procedure. The Environmental FMEAs must evaluate the opportunities for contamination to the process solution and final filled product container through any openings or apertures during manufacture. Items to consider when creating and updating Environmental FMEAs should include the following:

- Facility layout
- Clean room equipment, container, and tools
- Process materials (e.g. raw materials, components, container, and closures)
- Product pathway (formulation tanks, solution transmission systems, open versus closed systems):
  - i. Tanks
  - ii. Piping (flexible hoses), air flush, vacuum, or gas systems
  - iii. Agitators
  - iv. Pumps
  - v. Valves/Gaskets
  - vi. Change boards
  - vii. Filter housings
- Filter integrity test stations
- Cleaning/sanitization efficacy
- Physical routes of product transfer and associated controls
- Utilities (compressed gases and solutions)
- Available relevant microbial data (e.g. bioburden and endotoxin data from routine monitoring and validation)
- Product-specific data (e.g. growth promotion characteristics)
- Known or anticipated locations of product contact
- Locations likely to have a high potential for microbial occurrence
- Operations involving open or exposed product
- Critical interventions that are performed during the manufacturing process
- Locations where the product is exposed to the environment
- Line and process configuration
- Change rooms should be provided between clean rooms of different classifications. These rooms are intended to provide physical separation between areas of differing classification.
- Handwashing should be provided for personnel use prior to entry into change rooms. In lieu of handwashing, local sanitization procedures may be provided with appropriate validations.
- Entry of personnel and items for manufacture use (i.e. equipment, instruments, maintenance equipment, and tools) should be based on procedure to minimize and/or prevent the transfer of microbial hazards into areas of higher classification.

Materials such as paper and cardboard which shed particulates should be avoided. Wooden items (such as wooden pallets) that absorb and harbor microbial risk due to their porous properties shall not be permitted in the classified areas. Consideration shall be given to floors, ceilings, and walls relative to be able to be maintained in a "dry" state. These include but not limited to: coving, material of construction, cleanability, and resistance to agents used for maintaining their use. Procedures should be developed and routinely monitored for spillage response. These procedures should be aligned with the type and magnitude of the spillage to ensure removal to mitigate the risk of contamination/microbial transfer. All classified areas' air quality shall be qualified and maintained. Transitions through areas of different classifications shall be controlled so as to maintain the designated air classification for the activity occurring in the room into the transition area lands. For newly constructed or significantly redesigned facilities and equipment, consideration should be given to the ability for maintenance and/or repairs performed outside the critical areas – in particular outside of Grade A/ISO 5 and Grade B/ISO 7 areas.

5 Classified Areas

Monitoring of classified areas is continuous whether the room is in operation or not. Particles are measured through a continuous particle monitoring computerized system under measured conditions of temperature, pressure, and relative humidity. Microbial counts are sampled and read using settling plates both at the aseptic core and equipment surroundings inside the classified clean room. Locations of these plates and frequency of sampling are based on formal risk analysis, sampling requirements, and validation protocols. Alert and action limits are programmed based on classified room limits of operation and regulatory requirements.

Various technologies are employed in sterile manufacturing, such as isolator technologies, blow/fill/seal, aseptic customized tubing set as needed for product specifications, and terminally sterilized drug products, including sterile injectables (parenteral medicines). Protective personnel garments are specified for each clean room classification to cover hygiene, method of gowning, and required training and certification. General aspects of wall materials of construction, sinks, drains, changing rooms, airlocks (anterooms), and air supply and return circulation are major considerations in the design criteria and specifications used in clean room classifications. The use of sanitizers, disinfectants, detergents, and fumigation are closely controlled and specified for room and equipment cleaning.

The harmonization of regulatory requirements such as FDA and PIC/S determine acceptance criteria for limits during the processing of drugs (see Table 5.1).

Moist heat (saturated steam; not super-heated steam), radiation, or ethylene oxide gasification might be employed in sterilization of components, active ingredients, excipients, or drug products. Biological and chemical indicators are utilized to confirm sterility during routine production. In aseptic filling, liquid materials are filtered through a 0.22 µm sterile filter. Filter integrity is tested before and after use. Drug aseptic conditions require all filling operations to occur under class A (ISO 5 zone) conditions in the clean aseptic core; similarly, all capping will be under class A

TABLE 5.1
Media Fills Acceptance

Number of units filled	Contaminated units	Comments
<5000	0	No Contaminated units
<5000	1	Investigation/potential repeat of media fill
<5000	2	Investigation/required revalidation of media fill
>10000	1	Investigation
>10000	2	Investigation/required revalidation of media fill

conditions. Aseptically filled vials, cartridges, or prefilled syringes must pass 100% integrity testing.

To verify sterile products, sterility assurance level of 10^{-6} is achieved by overkill methods and sterility tests must confirm process capability. PIC/S conventions highlighted details for validation of aseptic processes and recommendations for sterility testing. In addition, it covered isolators for aseptic and sterility testing. Other guidelines governing these areas are EN ISO 14644-1, -2, ISPE guidelines, and US FDA aseptic processing guide.

6 Sterile Manufacturing Facilities

The focus of sterile manufacturing is on the state-of-the-art advanced aseptic techniques in tubing set design to cover challenging areas of aseptic manufacturing and demonstrate expert abilities to manufacture complex products that require borderline limits of process parameters. Strategic business and master planning in advanced aseptic facility design employ barrier technologies such as RABS, automated isolators that are equipped with CIP and SIP with full automation of product filling to mitigate the risk of contamination and increase the reliability of product integrity, and the use of single-use disposable technologies in view of managing global regulatory and compliance requirements.

The use of critical utilities in the manufacture of sterile products is fundamental to ensuring the sterility of that product. Water for injection (WFI), heating ventilation, and air conditioning (HVAC), clean steam (CSM), clean compressed air (CCA), and clean nitrogen (N2) are key utilities that a sterile manufacturing site will rely on in the component preparation, compounding, mixing, and filling operations of a sterile product.

Preservation of pharmaceutical product quality attributes is key in the selection of sterilization methods. Although terminal sterilization of finished product is always preferable, selection of an aseptic fill/finish route might be the choice to preserve the efficacy of API and compounding excipients. Accordingly, the selection of equipment in sterile manufacturing is determined based on product requirements to ensure efficacy and safety of the product. Irrespective of process selection, trending data of WFI and LAL-endotoxin is imperative to tracking sterility conditions and process reliability.

Due to potential contamination by personnel, mitigation of risk in sterile operations needs to focus on methods to minimize exposure and contact of sterile materials and components, and processing surfaces including instrumentation and equipment. In addition, recent developments in RABS and isolator technologies with advanced process control systems (PLC) automated controls are adding value to quality by design (QbD) and risk management in sterile manufacturing. The cost of such implementations remains a high price tag and should constantly be scrutinized as it relates to pharmaceutical product market penetration and value, and product handling specific requirements.

The design of clean rooms requires an integrated understanding of the cascading requirements from the aseptic core to the wrap-around corridors. Air filtration is channeled through a full planum HEPA system in ISO 5 zones to ensure unidirectional laminar flow over product filling and capping operations. Air flow uniformity control is verified through flow regimes and validated by smoke studies. Differential pressure cascading from high to low ensuring cleanliness of linked filling and lyoph-ilization suites, compounding suites, component preparation suites, aseptic corri-dors, and transfer rooms or pass-throughs for materials flow are all considerations in the design of sterile facilities to continuously establish conditions of effective bio-containment.

The main engineering design focus is to streamline process flow in terms of per-sonnel flow and material flow with the key criteria of mix prevention to avoid cross-contamination. Hence, line clearance, mix prevention, cleaning validation are major aspects of process design in sterile manufacturing. Pharmaceutical facilities selec-tion and construction require an understanding of materials of construction to ensure product life cycle. Therefore, QbD as described in ISPE guidelines, and quality risk management based on ICH Q8, Q9, and Q10 are important considerations in design criteria.

Sterile manufacturing requires strict control to mitigate the risks of potential contaminants such as particulates, microorganisms, and endotoxins. Cleanliness is key to ensuring that aseptic processing, or terminal sterilization is effective in controlling endotoxins levels and minimizing bioburden. There are specific chal-lenges to the various types of API or compounded products in processing liquids, colloids, suspensions, emulsions, powders, or semi-solids. Considerations regarding if the formulation supports microbial growth, high potency, toxic category, would require design provisions in terms of construction and installation requirements, and protection of product environment with the procedure to safely handle and store materials that could potentially harm personnel during manufacturing or in testing laboratories.

7 Personnel behaviors, Personal Protective Equipment (PPE), and Process Flow

Management and leadership strategies, mission, and guidance are key factors that will set the stage for personal behaviors. Indeed, this can be easily incorporated to cover a balanced scorecard approach to managing teams in these highly regulated manufacturing operations, namely:

* Define strategy and mission of the company
* Development of business processes that guide the interaction between the various disciplines engaged in aseptic manufacturing. Develop metrics and measurement of KPI including gap analysis
* Define technical knowledge management to cover compliance documentation and standard operating procedures (SOPs); procedures for the various departments and employees
* Develop a plan and program for training and growth to enhance individuals development with an emphasis on innovation

Sterile manufacturing SOPs should emphasize personnel gowning requirements, qualifications, and behaviors inside clean rooms and suites (ISO 5, 6, 7, & 8). Clean room design, barrier technologies, process design and qualifications, process control, environmental monitoring, data integrity, and production batch record accuracies should all be well understood and adhered to by following procedures to facilitate team integration and seamless operations in this regulated environment. ISO classifications are easily understood as it relates to particle counts in cascading clean rooms to reach the aseptic core. Training of personnel to understand these classifications and their importance in terms of various gowning levels will help personnel awareness in compliance to required procedures. Supervisory roles will focus on assuring conformance to written SOPs. In addition, personnel movements and monitoring will be observed in aseptic processing areas.

As companies evolve in moving from manual operations to automated ones, interventions during manufacturing operations will be reduced and this will lead to a reduction in deviations to justify these interventions. In this regard, unidirectional airflow patterns analysis needs to be confirmed due to operator interventions. Besides air handling unit (AHU) overall flow metered measurements, air flow velocity measurements should be taken in the proximity of HEPA filters and at the filling surface levels. Continuous Particle monitoring systems (CPMs) should be utilized with the aid of computers for non-viable particle counts. Air sampling should be through fixed cones close to the filling and capping station heads in sensitive locations under unidirectional airflow. Settling microbiological plates should be used to collect samples in sensitive locations to measure viable counts for the environmental monitoring program. Continuous monitoring of pressure differentials between the aseptic core (ISO 5) and the surrounding (ISO 6) cascading to ISO 7 should be balanced to ensure no pressure reversals as doors are opened to exit aseptic areas. Interlocking doors are important when doors are opened, the outward airflow in the airlocks should minimize potential ingress of contaminants. Therefore, the timing of open doors is critical and should be minimized.

8 Quality Approach for Systems Validation

Most Pharmaceutical or biological products are manufactured in batches with sterility assurance in mind. Batch size, processing stages and timing of these steps, fill weight and volume, type of container (vial, cartridge, PFS, BFS, ampoules, or ophthalmic bottles), line changes, line clearance, cleaning, disinfection, sanitization, and sterilization method requirements are all important factors in the setup of sterile operations. The impact of process steps on the product is of critical importance to ensure product performance. Science-based approach with all data measurements, risk mitigation methodologies and assessments (FMECA, HACCP, DOE), and validated systems are fundamental to product quality. Data integrity in every manufacturing step and laboratory testing is critical and non-negotiable to ensure product efficacy and safety.

Computer systems and computer-controlled equipment both in laboratories and manufacturing execution systems (MES) should be governed by computer systems validation (CSV) model as referenced in GAMP V. Risk mitigation exercise could be accomplished by using the 6M fishbone diagram (man, materials, machinery, methods, measurements, mother nature (environment)) to identify causes of failure and related effects or gap analysis in a particular process system. A failure modes and effects analysis (FMEA) can be accomplished by assembling a diverse team based on their areas of expertise. Empirical knowledge and experience are significant factors in identifying a measurable scale to quantify the risk level based on severity, probability of occurrence, and detectability (Scale = $S \times P \times D$). Like fault tree analysis (FTA, which is a deductive technique), FMEA's guide to mitigation and control options to minimize risk to the product is based on an inductive technique. By assigning grades to the causes of failure and multiplying these numbers to achieve an overall risk factor, a company procedure can assign levels of risk as critical, major, and minor and describe risk calculations in an SOP. This will show that the company's approach to QbD is incorporating risk management.

Sterile manufacturing and aseptic processing embody critical risk factors in process steps, which terminally sterilized products must achieve 6 log reductions in bioburden and ensure specified levels of endotoxins; aseptic processing must ensure sterile conditions and confirm container closure integrity. Aseptic processing techniques are primarily focused on the integrity of the filling and capping processes including the specific tubing set design to preserve aseptic transfer from bulk containers into individual vials, cartridges, or PFS. Sterile API transfers in the compounding and mixing stages including SIP of processing tanks, keeping formulations under aseptic conditions, and sterile filtration are all very sensitive and specialized techniques that are reflective of manufacturers' core competencies and typically are kept confidential. In addition, transfer to lyophilizers, filling and primary sealing, depyrogenation of containers, and closure in batch ovens or heat tunnels cover all components that need to be sterile as it meets a sterile product. The storage and transfer of sterilized equipment and components require specialized training of individual operators. Cleaning and validation of sterile process vessels and contact equipment require rigorous and validated methods for sterile and consistent operations.

Chemical and bio-contamination may transfer to products from the operating personnel, unclean materials, or contaminated equipment through cross-contamination from batch to batch or after the use of equipment with different products and without proper cleaning validation. Clean rooms might have higher counts of viable particulates that might be airborne and contaminate products. Cross-contamination can be manifested in residual products or cleaning agents that remain in difficult to clean crevices or surfaces. Enhanced cleaning and continuous training of personnel by using proper aseptic techniques will minimize all these risk factors.

Sterile manufacturing is conducted physically with documentation requirements governing all steps in the process to ensure compliance. Compliance in QA systems with respect to organizational structures, quality manual, quality procedures, change control, CAPA SOP, deviation SOP, master qualification plan, validation protocols, and reports, are all important quality systems requirements that are written in view of process performance qualifications (PPQ) and confirm to change control procedure, which might be computerized using CSV compliance packages such as TrackWise or GMPPRO. Production systems SOPs and batch record reviews are similarly controlled documents. Material systems governing sampling of received materials, storage conditions, and warehouse temperature mapping are all requirements to safeguard product quality. Laboratory controls and computerized data integrity are usually compliance audit targets to ensure method testing and validated methods relevance. Facilities and equipment that are used in sterile manufacturing are governed by guidelines, SOPs, and checklists. Change in any of these systems must be properly handled through change control and all activities must be conducted through written protocols and reports to clear a validated line to manufacture products or API. Drug products may be deemed adulterated even if no contamination is found. If the systems documentation or procedures are inadequate, the product might be considered non-sellable as validated methods are not in place to ensure the sterility of this product. It will be considered adulterated.

9 Dedicated Facilities

Facilities are specifically designed for certain products. For example, traditional non-toxic API products might be manufactured in one building. However, high potency oncology drugs are usually manufactured and tested in separate facilities. Similarly, penicillin antibiotics (beta-lactam antibiotic) are also manufactured in separate facilities to prevent cross-contamination. Hormone products are considered a specialty category and require separate manufacturing suites, and controlled substances (DEA Schedules I, II, and III) are strictly controlled in terms of receiving, storage, handling and manufacturing, and testing. This drug category requires separate physical paths and locations. These products might be prescribed for therapeutic purposes or as diagnostic standards in hospital and laboratory environments. Although these products might impose a significant risk or hazard if they contaminate other products, even at very low concentrations, they may have serious effects on some patients.

10 Contamination

Airborne non-viable particles are usually filtered through HEPA filters. If the contamination happens by the transport of particulates into the clean room by moving equipment into the room, people with improper gowning, or materials carried into the clean room without proper sanitization or handling such as contact parts that required sterilization before entry into the room, all these factors might jeopardize room cleanliness and create alarms on the continuous particle monitoring systems (CPMS), which create a deviation that requires justification for compliance reasons and continuity of sterile manufacturing in that suite. Vegetative, endospores, and molds are viable microorganisms that constitute a threat to room sterility and aseptic conditions. Extensive aseptic operator training and certification are very important steps to reduce the impact of people's unconventional movements in aseptic environments. In addition, the introduction of barriers between the surfaces of equipment used in filling and capping in the aseptic core from the operators surrounding the equipment is helpful in minimizing contamination. The introduction of more automation behind these barriers will improve the quality of consistent products and eliminate unnecessary operator interventions. Ensuring that unidirectional laminar flow confirmed by smoke studies is important in the qualification of aseptic core filling operations by confirming flow regimes through visual smoke. Operators enter through separate gowning and leave through de-gowning areas. The liquid is pumped through channels and filtered using a 0.22-micron sterile filter. All components entering the sterile core must be either depyrogenated or sterilized through irradiation or steam sterilization depending on component chemical characteristics to maintain its functionality and quality.

11 Containment

Handling of high potency ingredients, API, or compounded drug products might present a hazard or risk to the operator. Therefore, SOPs should govern the handling of products to minimize risk to their efficacy and prevent unsafe conditions. Similarly, the hazardous nature of API or compounds should be well identified, characterized, and graded with regard to exposure limits to protect the operator. This is usually achieved using containment chambers, such as fume hoods, bio-safety cabinets, glove boxes, or isolators. These systems are designed to protect the product and the individuals during handling of these toxic materials with very controlled specifications defining face velocities and recirculating compartment air through fitted HEPA filters or discharge to a bag-in/bag-out collection and filtration system for total discharge.

Isolators are used to create an aseptic environment in a sanitary zone to ensure biocontainment to protect the product at the same time operators are completely separated from the product or API. Isolators are cleaned and could be fully automated with CIP and SIP cycles. Nonetheless, more common is the use of vaporized hydrogen peroxide to decontaminate the inside of the isolator where the product is handled. To access a cleaned and aseptically prepared isolator, materials are introduced through ports (path-throughs) that are HEPA filtered to control ingress into the isolator.

Besides product manufacturing and transfer through connected isolators that are fully automated, isolators are used in the production of sterile media in biologics or microbiology laboratories. After media is formulated and sterilized, a docking isolator is used to transfer the media from the sterilizer into a transfer isolator and then into a half suit isolator for microbiological testing and evaluations. These isolators are usually decontaminated using vaporized hydrogen peroxide generators. All handling gloves attached to the various sections of the isolator train are tested for integrity as defined by operational SOP. Glove testers and visual inspection are clearly defined in SOP detailing glove testing steps to ensure integrity for reuse after sterilization. The background environment in rooms used for aseptic processing containing an isolator should be a minimum of ISO 8.

Milling and control of impurities in API are usually complex operations that require attention to details. Similarly, aseptic transfers of sterile API powders are also challenging. Therefore, an understanding of the design requirements to handle such operations is important in the design and preparation of aseptic facilities. Ample allowance for space and transfer systems is significant in ensuring the success of formulations utilizing such sensitive operations. Closed sterile vessels are used to transit through work areas. Aseptic interface valves are needed to transfer sterile API from holding containers into formulation tanks. Hence, aseptic transfers are the most challenging operations when producing aseptic products. Simulation media fills are necessary to prove the sterility of transfers and confirm overall process sterile capabilities.

With major emphasis on water for injection (WFI) water generation solution and powder aseptic transfers, other process validation operations present challenges related to advanced therapy medicinal products. Aseptic processing is constantly challenged to evaluate alternative transfer technologies as new product development might require novel ideas to be able to make them. Cell therapy scale-out has been an area of growth that requires specific planning. In addition, containment of cytotoxic compounds, automated visual inspection, formulation – filling of viral vectors, high-speed syringe filling expansions, robotic application in aseptic processing, combination drug delivery systems, and standardization of extractable testing are all areas that have been developing on fast track basis that require containment systems.

12 Suppliers

Equipment manufacturers such as ATS automation, Bosch, Brevetti, Marchesini, Innoscan, Korber Medipak, and Wilco have been developing new concepts in automated visual inspection with reference to Knapp studies. Robotics, AI, machine learning, deep learning, and interconnected workstations are being integrated into new installations. The Knapp test ties AVI results to characterized inspections by human operators to ensure product-specific qualifications and release.

The FDA CEDR and CEBR focus on Inspection of new product introductions in commercial production for drug and biologics pharmaceutical developments. Main emphasis is science-based methods that meet regulatory policy. Drug quality is managed through the offices of manufacturing quality and pharmaceutical manufacturing assessment governed by cGMP guidelines.

Aseptic processing's focus has been on providing flexible and reliable solutions to incorporate barrier and containment technologies. Many leading companies involve partnerships between vendors and pharma manufacturers, e.g. Gilead and Bausch & Strobel/Skan, Roche, Bosch, IMA, Teva, Fresenius Kabi, Evonik, Merck, and many others are developing practical tools to improve manufacturing capabilities and reliability. Case studies involve large- and small-scale facility and process design insights to meet product requirements and patient needs.

Contract Development and Manufacturing Organizations (CDMO) have been on the rise to assist in developing complex solutions in a short period of time, e.g. Vetter, Catalent, Cook, Patheon, AMRI, others, and subvendors. Diversification of projects in these organizations has created a cadre of experience that played a major role in the development of aseptic techniques. Alternate technologies such as options for decontamination of tubs, nests, and other prepared components in the area of regenerative therapeutic medicines utilize barrier technologies. The selection of a decontamination method is dependent on the log reduction requirement. Decontamination with pulsed light, E-Beam, hydrogen peroxide, and UV light are employed techniques that vary in intensity relative to microbial levels of reduction.

Aseptic track in viral vectors, CAR-T cell therapies, and protein production requires specific design implementations that are not conventional in aseptic sterile manufacturing approach. These individualized and personalized medicines implied smaller batch sizes. Besides maintaining high standards of aseptic fill-finish operations, these products are often live-cell-based or formulated in live-virus vector system, which presents a new set of challenges in maintaining aseptic transfers.

High potency cytotoxic compounds present other challenges in containment during aseptic processing. Regulatory and occupational safety, cleaning, and cross-contamination requirements for aseptic handling of high toxicity potent substances have specific requirements. Facilities and aseptic manufacturing lines must incorporate the protection of the operator and the product including robotic executions. Robots for transfer have been used with recent advancements in technologies. The applications include continuous particle monitoring systems in classified rooms environment and new ventures with innovative uses whereby products are transferred in isolators that are not fitted with gloves for intervention. This requires a modified and verified smooth process to capture the isolation technology benefits of added cleanliness, but reliable precise transfer and processing stations in component preparations, filling, and capping. Suppliers in these versatile areas have been developing technologies and systems to meet the aforementioned challenges.

13 Single-Use Technologies (SUT)

As cleaning validation is very demanding to prove that critical control points have been properly selected through DOE scientific methodologies and that the swabs are analytically evaluated to ensure systems cleanliness and repeatable steps are confirmed, SUTs are advancing sterile manufacturing techniques for product transfers into barrier systems. New risk factors must be analyzed to ensure SAL. Extractables and leachable compounds must be analyzed as part of materials compatibility. Small-scale and clinical batches have different requirements than large-scale (full-scale) economic manufacturing batches. The needs remain to fulfill aseptic quality standards irrespective of whether the batch is for clinical evaluation or commercial manufacturing. This covers requirements at compounding pharmacies, academic institutions involved in the making of test materials for clinical testing, or hospitals that are compounding preparation for internal use.

14 Master Qualification Plan

Validation program for process systems includes elements starting at the design phase of a project through decommissioning. The next figure shows the major elements of this program

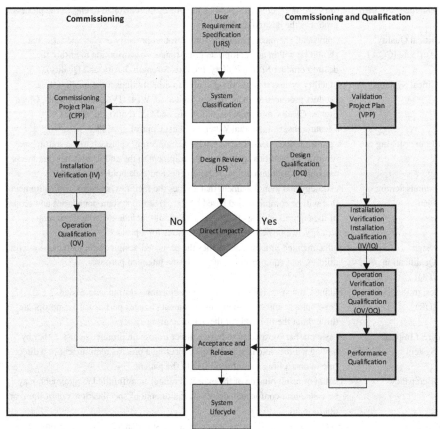

Definitions

Term	Definitions
Clean (Pure) Steam	Steam is used in pharmaceutical manufacturing operations that do not contain added substances and is controlled and monitored to specified quality attributes.
Compendial Water	Pharmaceutical water that meets the requirements and recommendations stated in applicable Pharmacopoeia documents.

(Continued)

Term	Definitions
Critical Component (CC)	Critical component is a component or equipment whose operation will affect product quality or regulatory compliance that can reasonably be expected to cause injury/safety risk to the patient/user if malfunction occurred.
Critical Equipment	An individual piece of equipment, which if it failed to meet its quality requirements, would have a direct impact on product quality.
Critical Instruments	Instruments that measure critical operating parameters.
Critical Process Parameters (CPP)	A process parameter whose variability has an impact on a critical quality attribute and therefore should be monitored or controlled to ensure the process produces the desired quality.
Critical Quality Attribute (CQA)	A physical, chemical, biological, or microbial property or characteristic that should be within an appropriate limit, range, or distribution to ensure the desired product SISPQ (Safety, Identity, Strength, Purity, and Quality).
Critical Systems	A facility system that has the potential to directly impact the quality of the product produced such as, but not limited to, WFI, HVAC-clean rooms, Clean Steam, Clean Compressed Air, Compressed Gas (Nitrogen, Argon), Cleanrooms, Compendial Water, and Pest Control systems.
Commissioning	A planned, documented, and managed engineering approach to the startup and turnover of facilities, systems, and equipment to the end user/owner that meets established user and design requirements and stakeholder expectations.
Commissioning Plan	A strategy and plan document that defines the facilities, systems, and equipment that will be commissioned based on the agreed-to system boundaries and scope of the commissioning effort. The plan should include the commissioning strategy, responsibilities, and expectations for a project.
Design Qualification (DQ)	A documented process that verifies the proposed design of the facilities, systems, utilities, and equipment is suitable for the intended purpose.
Design Review (DR)	Planned and systematic reviews of specifications, design, and design development and continuous improvement changes performed as appropriate throughout the life cycle of the manufacturing system.
Direct Impact System	A system that is expected to have a direct impact on product quality, either by having a direct role in supporting or meeting a process requirement or a direct role in controlling a significant risk to the patient.
Discrepancy	Datum or result outside of the expected range, an unfulfilled requirement, may be called non-conformity, defect, deviation, out of specification, out of limit, or out of trend.
FAT	Factory Acceptance Testing: Inspection and static and/or dynamic testing of systems or major components. Results of the FAT may be used to support the qualification of an equipment system conducted and documented at the supplier site in a manufacturer's setting prior to shipping to a user facility.
Good Engineering Practice (GEP)	Established Engineering methods and standards that are applied throughout the project and life cycle and ensure the effective satisfaction of stakeholder requirements.
Installation Qualification (IQ)	A documented process that verifies that the facilities, systems, utilities, and equipment, as installed or modified, comply with the approved design and the manufacturer's recommendations.

(Continued)

Term	Definitions
Installation Verification (IV)	A series of activities that demonstrate and document that installed facility, systems, equipment, and instruments comply with project installation specifications and vendor requirements prior to startup.
Impact Assessment	The process of evaluating the impact of the operating, controlling, alarming, and failure conditions of a system on the ability to meet process requirements and control risks to the patient.
No Impact Systems	A system that does not, either directly or indirectly, influence the capability of the manufacturing process to manufacture a product or has no effect on controlling risks to the patient.
Non-compendial Water	Water that meets the requirements of Drinking Water, but with additional treatment to meet specified process requirements
Operational Qualification (OQ)	A documented process that verifies that the facilities, systems, utilities, and equipment, as installed or modified, perform as intended throughout the anticipated operating ranges.
Operational Verification (OV)	The activity of reviewing, testing, checking, or otherwise establishing and documenting that a system operates according to written and pre-approved specifications throughout all specified operating ranges.
Performance Qualification (PQ)	A documented process that verifies that the facilities, systems, utilities, and equipment, as connected, can perform effectively and reproducibly, based on the approved process method and product specification.
Periodic Review	A documented assessment of documentation, procedures, records, and performance to ensure that facilities, equipment, and systems continue to be fit for purpose and satisfy regulatory compliance requirements. The frequency of periodic review is dependent upon the complexity, criticality, and rate of change.
Qualification	A process to demonstrate and document that the critical manufacturing facilities, systems, utilities, and equipment are suitable for the intended purpose.
SAT	Site Acceptance Testing: Inspection and static and/or dynamic testing of systems or major components. Results of the SAT may be used to support the qualification of an equipment system conducted and documented at the user facility.
Subject Matter Experts (SMEs)	Those individuals with specific expertise in an area or field.
System Owner	The person ultimately responsible for the availability, and support and maintenance, of a system and for the security of the data residing on that system.
Turn over Package (TOP)	Documentation packages generated by equipment manufacturers and contractors and submitted as equipment packages and systems are delivered or construction work is completed.
User Requirement Specification (URS)	A description of the requirements of the system in terms of product to be manufactured, required throughput and conditions in which the critical utility should be made.
Validation	Confirmation by examination and provision of objective evidence that the requirements for a specific intended use can be consistently fulfilled. (21CFR 820.3)

The intent of this section is to provide an understanding of integrated Commissioning and Qualification (C&Q) approach to make the process efficient and to standardize across facilities. The approach is based on a System Risk Assessment that identifies the critical quality attributes (CQA's), critical process parameters (CPP's) and design elements.

The C&Q approach demonstrates that the systems are suitable for the indented purpose and can be accepted and released for manufacturing use.

- Compendial water systems shall be validated per local procedures and documented through Validation Project Plans (VPPs).
- A change management program is required to capture changes as they are identified, assessed, and implemented.
- Good Documentation Practices are required in all Commissioning and Qualification process steps and other programs to demonstrate data integrity.
- Depending on the project size and complexity, the documents and deliverables may be combined. This should be documented in the C&Q Plan.
- SME's shall have enough knowledge of the processes and systems to ensure the systems are properly designed, tested, and operated as designed.

Validation details are provided on these major components of the validation program

- Regulatory requirements, Process and Business needs
- Specification and Design (User Requirement Specifications)
- System Classification
- Design Qualification
- Commissioning
- Qualification
- Acceptance and Release
- Maintenance
- Decommissioning
- Vendor Documentation
- Qualification Supporting Systems

The identification of Critical Quality Attributes and Critical Process Parameters during the initial design and process stage are key inputs to subsequent commissioning and qualification stages.

User Requirement Specification documents the details of the system's intended use as required by the user. This includes requirements related to CQAs and their process control strategy, other critical aspects requirements necessary to achieve product quality, as well as general requirements (for operational efficiency, safety, environmental controls, etc.). The set of requirements is an input to designing the system and will be identified and documented by appropriate Technical Subject Matter Expert/System Owners.

| Engineering Standards
Industry Standards
Business Needs
Quality Attributes
Regulatory Requirements
Process Parameters
Basis of Design
EHS | → | User Requirement
Specification
(URS) |

All utilities systems shall be classified according to Quality Impact to determine if qualification is required. Systems may be separated into sub-systems, which may have different classifications. The boundaries between these systems shall be clearly defined.

There are two methods for system classification, a System-Level Impact Assessment and a Commissioning and Qualification Program – Risk Assessment.

  Perform a System-Level Impact Assessment to determine and document if the critical utility system has direct or no impact to product or process.

  Perform a Commissioning and Qualification (C&Q) Program – Risk Assessment. If an object is identified to have potential impact to CQAs/CPPs and other process controls and hazards, then it is determined to have Critical Aspects. Utilities systems with Critical Aspects are Direct Impact Systems.

  The decision for Quality Impact shall be documented

  Manufacturing Systems with Direct Impact shall be qualified, have approved operational procedures, and a distinctive identification number or code. These systems shall also be assessed for inclusion in the Validation Maintenance Program.

  Commissioning and Qualification testing requirements are based on the System Classification:

Commissioning and Qualification based on System Classification

System Type	Commissioning	Qualification
Direct Impact	Required	Required
Direct Impact, Commercially available, off the shelf equipment	Not Required	Required
No Impact	Required	Not Required

System boundaries are markings on the design drawings used to distinguish systems from one another; this includes all components and piping to logically define what is included and what is not included within the system.

• System boundaries shall be defined and documented early in the project to enable development of the systems list and subsequent execution of system risk assessments.
• Systems are listed in the C&Q Plan

- Systems boundaries shall be defined such that the systems cover the entirety of the project scope
- Systems boundaries shall be documented in the process and instrumentation diagrams (P&IDs) or process flow diagrams (PFD).

RISK ASSESSMENT

The system risk assessment is the application of Quality Risk Management (QRM) to examine the product quality risk controls for direct impact systems. The assessment identifies the critical design and procedural controls that are required to mitigate system's risks to an acceptable level.

- System Risk Assessment shall be developed for direct impact systems.
- URS may need an update to ensure that the critical attributes and critical design elements have been included based on the risk assessment completion.
- Procedural controls elements shall be added to the commissioning and qualification scope
- Acceptance and release of the system shall be based on confirmation that the controls identified in the Risk Assessment are implemented as:
  - Design Controls have been shown to operate as required
  - Procedural controls are in place

DESIGN REVIEW/DESIGN QUALIFICATION

The Engineering Design Team shall employ Design Review procedures and practices to ensure that user requirements are incorporated into the design. The engineering design team shall ensure design elements exist to realize the Critical Aspects of the design – those elements necessary to control risks to product quality and patient safety or to achieve process user requirements. Design Qualification is not required for systems without Critical Aspects identified.

Formal Design Qualification may occur once the design has matured to the point where all quality risks have been controlled, process user requirements are included in the design, and Critical Aspects with acceptance criteria are identified and documented. The Design Qualification (DQ) examines the design of the manufacturing systems and identifies the Critical Aspects necessary to meet the process user requirements, control risks to product quality, adapt to variability impacting product quality, and meet cGMP and cGDP requirements.

Quality Validation Unit usually approves completion of the DQs.

Commercially available off-the-shelf manufacturing systems do not require a separate URS or DQ. However, qualification acceptance criteria of these manufacturing systems should be based on product and process requirements and control of any risks to product quality or safety.

- Design review shall include both quality critical and non-critical aspects of the system design.

- The project team shall define the method for conducting DR according to the system type, size, and risk. The method shall be documented and approved in the Project Execution Plan (PEP), Commissioning Project Plan.
- The project team shall have appropriate methods for recording and distribution of the results of the Design Review and any consequence changes.
- Project team is responsible for communicating to management any issues affecting design.

Design Qualification is performed only for direct impact systems. DQ requires more documentation, with Quality approval, to show traceability of the critical attributes and relevant critical design elements in the design to the CPPs and CQAs.

- DQ is performed through one or more design review(s).
- DQ shall verify that all business, quality, and regulatory requirements in the URS have been incorporated into the design.
- A final DQ report shall be issued stating the design is suitable for the intended purpose; the report shall be approved by Engineering, System Owner, and Quality.

COMMISSIONING AND QUALIFICATION

Critical utilities systems shall be deemed suitable for its intended use if it meets these criteria:

- It provides the ability to meet applicable process user requirements
- It provides the ability to control quality risks associated with the unit operation.
- It provides the ability to adjust the process control based on raw material and in-process variability.
- By extension, it does not impart a quality risk.
- It meets other GMP requirements as defined by relevant project documents and regulatory requirements.

DOCUMENTATION

Protocols shall describe how to conduct the testing for the manufacturing system including the type of testing being conducted, methods used, number of runs, and steps associated with critical aspects. If a grouping strategy is used, the protocol shall describe and justify the strategy.

Protocols shall be approved prior to execution. The strategy to accept commissioning testing and vendor documentation to satisfy qualification and/or validation requirements must be documented and approved by Quality Validation prior to execution. Protocols shall contain plans for any pre-requisites between validation activities. Engineering Validation and Technical Subject Matter Expert/System Owner will approve all protocols. Quality Validation shall approve all protocols that contain testing of critical aspects.

Where protocols and other documentation are supplied by a third-party providing validation services, appropriate User personnel shall confirm suitability and compliance with internal procedures before approval. Vendor protocols may also be supplemented by additional documentation/test protocols before use.

Critical aspects that do not meet specifications or established acceptance criteria shall be documented as a Discrepancy. Discrepancies shall be investigated to determine the cause of the failure, and appropriate corrective actions implemented prior to further testing related to the Discrepancy. For Discrepancies with known root causes, further testing can be completed after corrective actions have been appropriately implemented. A formal evaluation of the impact of the discrepancy shall be performed and documented, along with the corrective action.

Completed documentation including protocols and reports shall be reviewed by a designated SME.

The results of completed qualification protocols against the acceptance criteria, and any discrepancy resolution documentation, shall be summarized in a Qualification Summary Report and submitted for formal approval. Engineering Validation, Technical Subject Matter Expert/System Owner and Quality Validation shall review the final qualification package and approve this report.

COMMISSIONING PROCESS

Commissioning includes the controlled start-up, work instructions, regulation and adjustment, Shakedown, and other actions necessary to bring the critical utility system to a fully operational state. Commissioning will document the installation, operation, and performance testing for user requirements, Critical Aspects and other important engineering specifications.

The requirements for no impact systems shall include, at a minimum, operations procedures or manuals, maintenance procedures or plans, and documentation (logs, for example).

Where planned Commissioning testing includes critical aspects, these tests shall be approved by Quality Validation and completed prior to the start of Qualification.

- Commissioning activities shall follow Good Engineering Practices (GEP).
- A Commissioning Project Plan should be written to include:
  - Roles and responsibilities
  - Scope and strategy of the commissioning activities
  - Boundaries of systems to be commissioned
  - Documenting deliverables
  - Deviation handling
  - Startup and shakedown activities
  - Approach for system release for use.
- Activities shall be characterized by the application of qualification practices, including pre-approved protocols with acceptance criteria, follow good documentation practices, deviation handling, and reporting with appropriate approvals.

- Activities should include FAT and/or SAT, Pre-Delivery Inspections (PDI), Installation and Operational protocols (IV/OV).
- Protocols shall include a closure memo summarizing execution and status of deviations and/or pending activities.
- Protocol minimum approvals shall include system SME and system Owner.
- Separate protocols can be developed for automation and control systems depending on the level of complexity of the system.

QUALIFICATION PROCESS

Qualification is the documented verification that manufacturing systems and processes are properly installed and are suitable for their intended use. Qualification includes Installation, Operational and Performance Qualification (as required). Qualification confirms that (a) critical aspects have been verified to meet the pre-determined acceptance criteria; (b) process user requirements that affect product quality are satisfied.

The IQ/OQ includes verification of the installation and operation of critical aspects against pre-defined acceptance criteria. To execute the IQ/OQ protocol, the documented commissioning and other verification work are reviewed and/or testing completed to document evidence that each critical aspect has met its acceptance criteria. The reference location of this evidence is noted in the IQ/OQ protocol. If either (a) the equipment is simple such that commissioning is not warranted or (b) additional final inspections and/or operational testing is warranted under controlled conditions, then IQ/OQ protocols shall be used to control and document such activities. Instruments used for testing must be calibrated.

Some systems containing critical aspects may require performance qualification (PQ) to demonstrate the critical utility system or systems acting together consistently meet CQAs and CPPs including conditions for testing of upper and lower limits as applicable. Assays used during PQ studies must either be compendial methods or validated prior to use.

The decision to proceed with PQ (if required) should be based on a review of the IQ/OQ to confirm that all critical aspects and other items are in place that may be needed to perform PQ testing.

A Validation Project Plan should be written to include:

- Roles and responsibilities
- Scope and strategy of the commissioning and qualification activities
- Boundaries of systems to be qualified
- Documents deliverables
- Deviation handling
- Startup and shakedown activities
- Approach for system release for use.
- Qualification practices and documents shall include pre-approved protocols with acceptance criteria, follow good documentation practices, deviation handling and reporting with appropriate approvals.
- Activities shall include Commissioning (IV/OV) in addition to IQ/OQ/PQ protocols.

- Commissioning and Qualification protocols can be combined into IQ/OQ/PQ protocols subject to the project scope and complexity of the system.
- Protocols shall include a closure memo summarizing execution and status of deviations and/or pending activities.
- Protocol minimum approvals shall include system SME, Quality/Validation and system Owner.
- Separate IOV/IOQ protocols can be developed for automation and control systems depending on the level of complexity of the system.

VENDOR DOCUMENTATION

Qualification requirements may be supported and/or met by documentation provided by the vendor. This includes but is not limited to material certification, as built P&IDs, operational manuals, spare parts lists, and preventative maintenance and calibration procedures. The vendor documentation supporting qualification will be reviewed and accepted as part of the validation documentation.

QUALIFICATION SUPPORTING SYSTEMS

There are several quality systems that must be established in order to operate qualified systems meeting GMP regulations. These supporting systems required for each critical utility include:

- Standard operating procedures
- Preventive maintenance program
- Instrument calibration schedules
- Operator training and qualification
- Engineering files (drawings, manuals, spare parts lists) established

The completion of these Qualification Supporting Systems shall be verified and summarized within validation protocols and reports or other designated quality approved document (i.e. Qualification Summary Report, Acceptance and Release Report, Change Control) prior to release for manufacturing operations.

ACCEPTANCE AND RELEASE

A critical utility system or functional area shall be released for manufacturing operations through approval of the qualification testing, where the critical utility system is deemed to be suitable for their intended use. This may be completed as part of the Qualification Summary Report including the Qualification Supporting Systems, or through the grouping of multiple reports (i.e. Installation/ Operational Qualification Report per system, Performance Qualification Report for grouping of systems, Acceptance and Release Report for grouping of systems) where the Acceptance and Release report includes the Qualification Supporting Systems.

A formal release for the next stage of validation (Process Performance Qualification) will be authorized by Quality Validation. When a conditional approval is required, a documented rationale which includes assessment of risk, justification, and method for tracking actions to completion, is required. Site Quality Validation and Engineering will formally review and either accept or reject the request. The conditional approval will be documented and included in the Validation report.

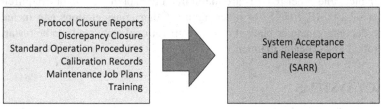

MAINTENANCE

The performance of critical utility systems shall be periodically evaluated to confirm they remain in a validated state. Validation of Maintenance is a defined set of activities including, but not limited to, monitoring, review, and setting the minimum requirements to periodically verify that a validated system has remained in a state of control.

For critical utilities systems, a risk-based approach is recommended to evaluate the need to conduct Periodic Reviews based on regular reviews of quality attributes, reporting of performance trends, maintenance activities and calibration metrics, and checks on system drawings and component tagging.

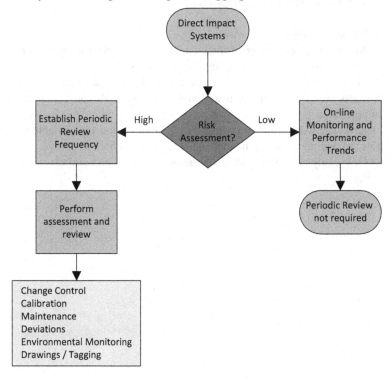

DECOMMISSIONING

Critical utility systems that are to be removed from service shall be decommissioned. At a minimum, decommissioning shall ensure that necessary operating, maintenance, and other records are maintained to support regulatory requirements, whether of a GMP, EH&S, or other nature. Other considerations can include the need to decontaminate or place it in a non-operational state ready to be restarted in a controlled and efficient manner.

LEGACY SYSTEMS

For existing facilities with qualified critical utilities systems, the supporting documents that were used to demonstrate suitability for use may have followed a different approach. This should not be a concern if the respective documents conform to the procedures that were effective at the time. Nonetheless, critical utilities that require materials specifications upgrades due to changes in regulations should develop a set of compliance documentation that confirm validation of these system modernization or upgrades (Remediation). However, instead of retrospectively generating specifications and design phase deliverables, the focus of a more efficient C&Q approach should be on identifying business/process user requirements and assessing risks to the system quality requirements. If the system documentation requirements are lacking, then URS and validation legacy protocols might be developed to properly describe and document the existing system.

- Define system user requirements
- Perform a risk assessment of the current system against identified critical process parameters and critical quality attributes
- Verify that critical process parameters, critical quality attributes, and other business needs are properly tested and measured, calibrated, maintained, and operated
- Ensure system drawings and equipment documentation are available to support the operation, maintenance, and change controls.
- Review equipment and system operational history and address recurring issues.

15 Water For Injection (WFI)

User Requirements Specification (URS) for WFI Distribution System considers the normal range of equipment operation. The purpose is to specify user requirements for equipment that are to be used in supporting the production of finished pharmaceuticals in a cGMP environment.

URS serves as the input to subsequent project risk assessment activities, design objectives, control strategy, and acceptance criteria for testing and qualification of the manufacturing system.

1- Materials of construction of WFI loops will be SS 316L
2- All piping weld connections must be boroscope tested at 100% of all welds. Visual inspection of welds to determine finish quality must be random and cover at least 10% of the welds.

Compliance Determination

Regulatory Body	Number	Title
FDA	21 CFR parts 11	Electronic Records and Signatures
FDA	21 CFR part 210	cGMP in Manufacturing, Processing, Packaging, or Holding of Drugs
FDA	21 CFR part 211	Current Good Manufacturing Practice for Finished Pharmaceuticals
FDA	21 CFR part 820	Quality System Regulation

Engineering Standards/Guidelines

Regulatory Body	Standard/Guide
FDA	High Purity Water System Inspection Guidance (7/93)
USP	<85> Bacterial Endotoxins Test, <643> Total Organic Carbon <645> Water Conductivity, and <1231> Water for Pharmaceutical Purposes v.40
ASME	BPE 2019
ISPE	Approaches to Commissioning and Qualification of Pharmaceutical Water and Steam Systems, Commissioning and Qualification, Applied Risk Management for Commissioning and Qualification, Water and Steam Systems
ISO	22519 2019
GAMP 5	Good Automated Manufacturing Practices
ASTM E2500	Standard Guide for Specification, Design, and Verification of Pharmaceutical and Biopharmaceutical Manufacturing Systems and Equipment

Key Abbreviations and Terms

Abbreviation / Term	Definition
CPP	Critical Process Parameter
CQA	Critical Quality Attribute
DW	Distilled Water
FDA	Food & Drug Administration
FDS	Functional Design Specification
OSHA	Occupational Safety Health Administration
P&ID	Piping & Instrument Diagram
TOC	Total Organic Carbon
URS	User Requirement Specification
USP	United States Pharmacopeia
WFI	Water for Injection

PROCESS DESCRIPTION

WFI is generated utilizing pre-treatment system and distillation units and is then stored in silos (storage tanks). Water is recirculated via primary pump skids through primary loops. Secondary pump skids, pump WFI through secondary recirculating loops to use points. Pump skids provide enough flow to assure adequate volume for operations, as well as maintenance of minimum Reynolds' number to assure turbulent flow on both supply and return legs of the loops. The skids also utilize heat exchangers to accommodate periodic sanitization. Silos and tanks will each be equipped with a vent pipeline with two 0.22 μm filters allowing pressure equalization when the water level changes. Typically, WFI water temperature is kept at 80°C in aseptic operations, but the temperature is usually reduced using heat exchangers to meet point-of-use requirements for operator handling and product requirements conditions.

SPECIFICATIONS

LIFECYCLE REQUIREMENTS

Equipment Suppliers of the silos, tanks, equipment, piping, instrumentation, etc., are expected to provide for manufacturing cycle:

- Written communication of any deviations from User specifications.
- Written communication of changes to the design specification.
- Manufacturing/completion project schedule.

PRODUCT AND PROCESS USER REQUIREMENTS

PRODUCT DESCRIPTION

WFI Distribution System distributes WFI throughout the facility. WFI distribution system transports water for injection primarily to the mix tanks to become compounded solutions when API and excipients are added.

WFI Critical Quality Attributes (CQA)

Process Step Reference	CQA	Specification
Distribution	Micro	≤ 10 cfu/100 mL (Absence of USP indicator pathogenic organisms)
Distribution	Endotoxin	< 0.25 EU/mL
Distribution	TOC	≤ 500 ppb
Distribution	Conductivity	≤ 1.3 µS/cm @ 25°C

Critical Design Aspects (CDA)

Process Step Reference	CDA	Associated CQA	Specification
Distribution	Materials of Construction	Micro, Endotoxin, TOC, Conductivity	316/316L or USP Class VI Elastomers
Distribution	Surface Finish	Micro, Endotoxin, TOC	≤ 30 Ra
Distribution	Slope	Micro, Endotoxin, TOC	0.5%
Sanitization	Absence of Dead Legs	Micro, Endotoxin, TOC	None with L/D ≥ 2.0*
Sanitization	Drain ability	Micro, Endotoxin, TOC	Low point drains present at low points

*Dead legs should be avoided wherever possible. If it cannot be avoided due to design constraints or other means, examine the risk of dead leg section with L/D ≥ 2.0 and determine risk acceptability on a case-by-case basis including manual ability to drain that section.

Critical Component (CC)

Process Step Reference	CC	Associated CQA	Specification
Distribution	Vent filter	Micro, Endotoxin, TOC	0.22 microns

Process Parameter Requirements

Process Step Reference	CPP	Operating Range	Associated CQA
Distribution	Flow	Reynolds number ≥ 2500	Micro, Endotoxin, TOC
Sanitization	Temperature	≥ 80°C @ coldest or worst-case point	Micro, Endotoxin, TOC
Sanitization	Time	TBD in Validation (capability NLT 8 minutes)	Micro, Endotoxin, TOC

Critical Instrument (CI)

Process Step Reference	CI	Associated CPP	Associated CQA
Sanitization	Temperature transmitter	Temperature	Micro, Endotoxin, TOC
Sanitization	PLC clock	Time	Micro, Endotoxin, TOC
Distribution	TOC analyzer	N/A	TOC
Distribution	Conductivity probe	N/A	Conductivity
Distribution	Flowmeter	Flow	Micro, Endotoxin, TOC
Distribution	Pressure transducer	Pressure	N/A

OPERATING RANGES

WFI system shall be a recirculating loop-based design with any sections of piping Length/Diameter $\geq$ 2:1 pre-approved by the quality unit. WFI Distribution System shall have a validated process control system. WFI system shall be capable of monitoring, in real time, line pressure, temperature, TOC, flow, and conductivity at a minimum.

Non-pressurized vessels in uncontrolled environments shall be equipped with a vent filter for bi-directional flow of make-up air. All points-of-use shall contain a means of collecting a sample aseptically and without adversely affecting the environment. All loops shall contain at least one means of collecting a representative sample for each loop. Rinse assemblies will be automated to decrease moisture generation in controlled environments.

Each point-of-use/product contact valve shall be sanitary diaphragm e.g. zero static design, double block-and-bleed or mix-proof valves. Heat exchangers will be sized to heat water in each circuit to sanitization temperature in <= 30 minutes. Pumps will be controlled by VFD and pump motors must be VFD rated, high efficiency, totally enclosed fan cooled (TEFC). In addition, All equipment operating ranges must be clearly communicated and controlled within the system manufacturer's recommended safe operating range.

Recommended spare parts list must be provided and spare parts are required to be on-site prior to system startup. Most point-of-use valves shall be automated and controlled for operation. Manual POU shall be placed throughout the system to allow for a tap-in for cleaning where applicable. Modulating valves shall be used to control flow into tanks to achieve target weights. The recirculating loops in the distribution system shall be capable of being hot water sanitized in a USP 1231-compliant manner. The system will heat WFI at user-defined temperatures for process requirements.

MATERIAL OF CONSTRUCTION

Materials of Construction requirement for all components that have solution contact:
Components include pipes, fittings, instruments, valves, tanks, pumps, heat exchangers, etc.

METALLICS:

All metallic materials of construction must be 316L stainless steel and have a maximum internal surface finish of 0.6 micron/25 micro-inch Ra.

PLASTICS:

All plastic materials shall be Polyvinylidene Fluoride (PVDF) or Polypropylene (PP)

ELASTOMERS:

All elastomers and alternate materials shall be capable of withstanding system temperature and chemical exposure and be constructed of Ethylene Propylene Diene Monomer (EPDM), FKM Viton, Teflon/Polytetrafluoroethylene (PTFE) or Platinum Silicon.

GLASS:

Tempered Glass, such as sight glass, shall be homogeneous, scratch-proof, impervious, inert, and non-toxic.

Material traceability will be provided with every metallic component that will have solution contact. Each pipe component should be marked with an indelible identifier that can be traced back to the original material certificate. Identifiers can be Heat, Lot, or a reference number linked to the Weld Record. Instruments, valves, and hardware are to be marked or tagged. Fittings will be identified in the weld map with traceability to the materials of construction and internal surface finish. Examples of acceptable material certificates:

- Certified Material Test Report
- Certificate of Conformance
- Cut Sheet

A surface finish verification certification shall be provided for each component that will have solution contact, along with a certificate of calibration of the profilometer. The surface of each metallic component that comes into contact with the product must be passivated in accordance with SOP. Passivation must have official documentation and the passivation numbers shall each be linked to the corresponding components.

Exterior surfaces of components that reside in a controlled environment are preferred to be 316L. 316 or 304-series stainless

WELDING REQUIREMENTS

Piping shall be orbital welded. Welding meets ASME BPE 2019 for Bioprocessing Equipment. Purge gas is ≥99.99% pure argon. Pressure Vessels shall be welded in accordance to appropriate regulatory agency requirements and ASME. Welders and weld inspectors must be qualified for weld procedures. Welder and Inspector welding program credentials must be furnished to project manager prior to performing any welding job.

Welds on solution piping must be passivated. The passivation must be documented and included with the welding documentation package. These welds may be spot-passivated if reachable. Non-accessible welds must be bath or circuit passivated. Weld repairs shall be permissible during initial fabrication if caused by a lack of penetration, lack of fusion, or unconsumed tack welds. Defective welds that cannot be repaired shall be cut out. Welding procedures shall include:

- Cleaning procedure
- Weld repairs
- Visual Inspection of the external of all welds
- Visual inspection $\geq$10% of internal portion of welds. The percentage of inspection shall increase with each failed weld.
- Visual inspection of all manual welds
- Inspection documentation
- Unique weld numbers
- Date performed
- Welder identification
- Description of materials being joined
- Heat numbers

Verification of all welds will be 100% done through borescope videos. Weld documentation/Weld Map (drawing of the area/components) including:

a. Weld location identification and reference numbers
b. Welder signature and date at end of weld job
c. Weld inspector signature and date at end of weld job

1) Weld Record (table with the following columns):
   a.   Weld Map reference number
   b.   Weld type – manual/orbital/etc.
   c.   Component #1 material and reference to the traceability document (if solution contact)
   d.   Component #2 material and reference to the traceability document (if solution contact)
   e.   Rod/Wire material and reference to the traceability document (if solution contact)
   f.   Performed by/date
   g.   Inspected by/date

Note: Welder and Inspector welding program credentials must be on-file. This documentation must be completed immediately following any welding activity (daily, at a minimum) and shared with the project manager on a weekly basis for review.

Welding symbols should be in accordance with AWS A2.4 or other equivalent ASME standard.

CONSTRUCTION REQUIREMENTS

Insulation will be utilized for piping anything within 7 feet of the working floor or platform (following OSHA standards). Equipment insulation must be non-particulate generating and shall be jacketed per OSHA requirements. WFI storage vessels will have vent lines and be protected by vent filters (electrically heat jacketed where applicable). WFI Silos and Tanks Vent filter and Housings shall be sized to accommodate the design flow rate. Storage vessels shall be kept wetted via spray balls.

Distribution system piping shall be appropriately supported and sloped at a minimum of 0.5%. Slope measurements should be made between pipe hangers/supports and at each change of direction. All system low points shall have a low point drain. All piping shall be orbital welded. Orbital welded connections are required for product contact piping. All welds will meet SOP requirements and surface finish requirements.

All piping will be clearly labeled for content and direction of flow. All Points of use (POU), silos, loops, subloops, and buffer tanks, shall be clearly labeled and identifiable. Each vessel shall have a manway with a minimum opening diameter of 22 inches. Passivation procedure and reports of all 316/316L components being passivated and absence of passivation chemicals post-passivation. Manufacturer's drawings shall be provided for all components, piping excluded. Hydrostatic/leak test reports shall be provided for all components. All attempts will be made that all pump head will be directly after the pump. If this cannot be accomplished, appropriate engineering controls must be in place to ensure adequate turbulent flow and performance of the system. The vendor shall ensure that loop can operate with either one of a redundant pump system.

LUBRICATION REQUIREMENTS

All lubricants used must be detailed in the design documentation. Include lubricants used in the mechanical aspects of the equipment as well as the lubricants used in the product assembly. Where direct lubrication is required, the design and construction must be such that the lubrication cannot leak or drain into the product or into the product contact areas. Where pneumatic machine functions require the use of lubricated compressed air, only USP and FDA-approved lubricants are acceptable: e.g. White Mineral Oil, USP

Lubricants with USDA and FDA approval and "AA" classification shall be used where incidental contact may occur.

ELECTRICAL REQUIREMENTS

Electrical systems must be designed to applicable Electrical and Controls Specification. Where applicable, "as built" instrumentation wiring diagrams shall also be provided. Electrical panel penetrations should be made from the bottom of the panel when possible. Reference NFPA 79, OSHA, and User Standards for additional Electrical Requirements.

All electrical panels need to have externally mounted disconnects for Arc Flash compliance. Panels in controlled environments are to be NEMA 4x; all others a minimum of 3R rating. All electrical panel connections will be IR mapped and tested periodically to identify hot spots and should be corrected to prevent ignition and fire hazard. High voltage cabinets should be UL listed.

Safety Requirements

All pinch points and hazardous areas must be guarded per OSHA standards using required materials of construction. E-stop buttons must be located close to operator work areas and must release all potential energy immediately upon engagement. Restarting the equipment shall require a reset and disengaging of the E-stop. All access doors must have interlocks as described in Electrical and Controls Specification. Access doors must be interlocked with magnetic mechanisms to not allow entry to the machine while it is running.

The design must permit the electrical, pneumatic, and hydraulic devices to be locked out during maintenance and repair activities. All lockout/tagout points must be easily accessible. The control system must be designed to operate in a fail-safe manner and meet OSHA safety requirements. The safety circuit shall be at least category 3. Jog capability is acceptable only with a single momentary switch. Sharp corners and edges must be eliminated. Safety circuits must not be able to be bypassed. Electrical panels with >50v are to be designed with incident energy below 1.2 Cal/cm². Fixed mount HMIs are to be set at a height of 54" from the floor to the center of the screen. Adjustable mounts are to cover the range of 54"–68" to the center of the screen. All system components and connections (pumps, heat exchangers, instruments, POU valves, tank connections, etc.) must be easily, safely, and ergonomically accessible.

All heat exchangers shall be, on the shell side, protected by an appropriate pressure relief device where required. Distribution piping in contact with hot WFI and readily accessible by personnel shall be insulated for safety. Each storage vessel shall be equipped with a certified rupture disc.

All equipment shall be appropriately guarded. Required EHS checklists will be completed at the appropriate time in the project life cycle. All equipment with potential electric discharge must be grounded and bonded to code.

Operational Requirements

Process Constraints and Limitations

WFI velocity in the piping system shall be limited to no more than 10ft/s.

Process Control System

Vessel volume, liquid additions, temperature monitoring, quality monitoring capability, and sanitization shall be interfaced with and controlled by an automation system. Vessel systems shall be capable of being monitored and/or controlled from a local HMI within controlled space interfaced to an automation system.

Electrical components shall be mounted in NEMA-rated enclosures suitable for the local environment and service. Control panel assembly and electrical motors and systems shall be UL approved.

- Loops have low and high supply pressure alarms
- Loops have low and high return pressure alarms
- Tanks have low- and high-level alarms
- Loops have Hi TOC and Conductivity alarms

The rupture disks in the WFI system will generate an alarm if triggered. All instruments shall be suitable for the local environment and service. There shall be no passwords that reside in programming code and no protected source code. All electrical panels that contain controllers must have externally mounted communications ports with 120VAC receptacle for programming and servicing.

OPERATOR INTERFACE

Display measurements in inches. Display time using 24-hour clock. Display speed in gallons per minute.

Control system must be able to generate accurate and complete copies of records in both human readable and electronic forms for inspection, review, and copying. Control system must be able to discern invalid or altered records or entries. Control system must be able to protect records to enable their accurate and ready retrieval throughout the records' retention period. Control system must be able to ensure that system access is limited to authorized individuals where:

- System access is restricted based on user responsibilities and application functions.
- System user access levels are documented.
- Password expiration rules exist for the system.

Control system must have authority checks to ensure that only authorized individuals can:

- Access and Use the System
- Alter a record
- Perform the specified operation

Control system must provide audit trails for all objects and users that must:

- Be secure
- Be computer generated
- Be time and date stamped (e.g., server time)

Audit trails must independently record the date/time of operator entries and actions that:

- Create electronic records
- Modify electronic records
- Maintain electronic records
- Delete electronic records

Audit trail must record changes and must not obscure previously recorded information. Control system must ensure the audit trail is secure from being modified/deleted and cannot be disabled. Control system must ensure that the audit trail documentation must be retained for a period at least as long 90 days and must be available for review and printing.

POWER LOSS AND RECOVERY

On power restoration, the system shall not restart without operator or communication-link input. No damage to machine will occur as a result of going to the safe state. Integrator will supply instructions for recovery from catastrophic control system failures. It is generally accepted that the system shall protect in the following priority: Personnel, Equipment, and then Product. Machine will be manually restarted based on operator inputs (Operator initiates power recovery sequence).

CLEANING AND SANITIZING

All components that will contact solution will be initially cleaned by the passivation requirement. Further cleaning will be accomplished when the system comes on-line through sanitization requirement

MAINTENANCE REQUIREMENTS

Recommended Maintenance procedures will be available prior to completion of FAT. Original equipment manufacturer part numbers are required for all spare parts. Spare parts shall be described as either "critical" or "general". Drawings will be provided in a format agreed to by User Engineering for all machined parts.

The Operations Manual will include a section on troubleshooting and repair and preventive maintenance.

TRAINING AND DOCUMENTATION REQUIREMENTS

Material cut sheets or specs are required for all WFI connections. Training documents will be supplied for Operation and Maintenance procedures. Operator training will be available during commissioning, potentially on multiple shifts. Maintenance specific training will be available during commissioning, potentially on multiple shifts. A draft Operations Manual will be available prior to FAT and the final version at FAT. Operations and Maintenance Manuals and training will be provided prior to system go-live. Start-up and tuning instructions (water flow quantities, correct instrument readings, etc.) shall be provided in these manuals. Design documents, parameters, data sheets, technical specifications, P&IDs, and drawing files will be provided and will include the requirements for preventive maintenance.

16 Integrated Facility Design

In general connotations, the aseptic core where filling and capping occur should be controlled and classified to ISO 5 standards. Around the filling and capping and outside the RABS, ISO 6 standards might be applied. Outside the filling and capping suites, an aseptic corridor with ISO 6 classification is appropriate. All material and personnel flow adjacent rooms must be designed with dedicated paths for in and out. Materials can be transferred through path-through design. These adjacent rooms can usually cascade from ISO 6 to ISO 7 towards gowning rooms. Exit to ISO 8 aseptic corridor and out to CNC space (sanitary). All component preparation suites must be in ISO 7 classified rooms. All compounding and mixing of product formulations must be in ISO 7 classified suites.

The manufacturing process includes the cascading space described above, and other work-in-progress (WIP) areas for handling and storage or staging for product fill/finish. To list a few process steps, these will include weighing of excipients, API weighing and dosing, milling, mixing formulation to add solids to liquids, which might include gas purge to blanket product. Following compounding a sterile filtering might occur before transfer of sterilized materials and components such as stoppers or caps into the filling suite.

Lyophilization incorporate the filling and partial stoppering of vials before loading into lyophilizers, whereby water is removed from the product under specific vacuum conditions. This is followed by unloading sealing and capping, washing in case of high potency drugs, drying, labeling and tray off, then serialization and packaging.

A typical and efficient setup of a sterile drug manufacturing building in a two-level construction will typically house all the manufacturing suites on the first level including all necessary utilities, bathrooms, break rooms, PPE rooms and visual corridors for line observations. The second floor might be split into an equipment mezzanine to house Air Handling Units (AHU) for HVAC that support the clean rooms. Other critical utilities installations, control cabinets, and maintenance support space. Chemistry and microbiology laboratories might be on the second floor in support of manufacturing operations sampling testing requirements including environmental monitoring of clean space, and critical utilities measurements and testing to ensure efficient operations. A wrap-around sanitary corridor might provide for ease of access to all these labs and support equipment needed.

Manufacturing classified rooms are controlled through HEPA filtration and room pressurization, whereby the aseptic core has the highest pressure and cascades down to the wrap-around corridor. The intent is to mitigate risk by minimizing product contamination. The use of PPE protective clothing and separate gowning and de-gowning rooms plus the sterilization of all components needed in t,he fill room utilizing various pretreatment methods, is the recommended practice to support sterile transfers into the aseptic core.

Controlled not classified designation implies that the space is cleanable, but with access control and air is filtered for ventilation. Personnel transfers, and required garment is discretionary. No environmental monitoring is required, but may be optional to control ingress of mold.

Capping areas must be close and adjacent to filling, but with proper barriers to prevent particulate matter transfer into open containers. All design criteria and selection of capping equipment must emphasize capping and crimping systems that are designed to mitigate the shedding of particles. Additional measures such as vacuum suction systems might be employed very close to the capping head to ensure that any generated particles would be collected through the vacuum system.

Major advances in aseptic/sterile manufacturing focus on processes that allow for minimal personnel interventions through further automation and robotic manipulations. Furthermore, advances in gating and automated access tools in specific critical locations might allow for personnel interventions but with minimal intervention impact.

Drug Substance / API Processing	Drug Product /Fill Finish
Dispensing	Dispensing
Compounding	Compounding
Sterile filtering	Sterile filtering
Crystal seed	Container/cap preparations (depyrogenation/sterilization)
Crystallization	Aseptic transfer of components and materials
Drying	Aseptic filling/capping
Milling	Lyophilization – Partial stoppered vials transfer
Sizing	Complete capping and crimping
Container packing	Terminal sterilization if not aseptic
Closure	Manual or automated visual inspection
packaging	Serialization and labeling packaging

Aseptically processed products that cannot be sterile filtered, e.g. sterile powders must be dispensed in an Iso class 5 environment. Liquids can be metered or weighed. Vessels might be placed on load cells or floor balance to achieve desired weights. All necessary calibrations must be in place and paying attention to interference from auxiliary cables or hoses that might be attached to the equipment such that there is no impact on the accuracy of weighing.

Every precaution must be taken to prevent cross-contamination during dispensing. This could be due to the carryover of residual materials from a previous batch, or parallel dispensing of different products. Proper GMP implementations would minimize the risk of contamination. For example, Established validated cleaning procedure between batches is a clearance procedure that must be followed by trained operators to achieve correct, clean and efficient dispensing.

Double-door batch ovens, path-through steam sterilizers or chain and conveyor depyrogenation tunnels can directly transfer sterile components into the aseptic filling environment. If this type of sterile transfer is not available, components might be double wrapped in sterilizable sheets that permit air/steam ingress while maintaining sterility of contents. Components that are gamma irradiated or ethylene oxide gas sterilized can be aseptically transferred to the ISO 5 clean room. Precautions must be taken through validated steps to remove all disinfected outer wraps without introducing risk of biocontamination into the ISO 5 environment.

WFI is being used as the basic solvent for aqueous formulations including emulsions. All processing equipment must be sterilized in validated cycles. Liquid mixing tanks are SIP sterilized. Aseptic powder transfers are usually challenging and must be validated. In addition, appropriate expertise should be sought to prevent dust explosions, as the understanding of angle of repose, static electricity, grounding and bonding of equipment, powder density, void factors and moisture content effects on ignition/explosion in the presence of oxygen.

Critical process utilities systems such as WFI, CSM, CCA, Clean N2, and Clean air in classified areas are considered as direct impact to product quality and safety. These critical systems are designed to prevent product contamination in pharmaceutical sterile manufacturing. Materials of construction of all storage and distribution systems are very important to specify such that they will contribute to transport of fluid safely without any ingress of contaminants. Besides the chemical nature of materials, continuous cleaning and sanitization as validated systems require is a crucial aspect of maintainability against microorganism growth. Other process utilities support systems are important for operations reliability. Critical utilities must not be reactive or absorptive and must withstand chemical and sanitization repeated cleaning.

The impact of power failure in sterile manufacturing aseptic facilities must be of major design consideration. All critical controls must be backed up with Uninterrupted power supply units (UPS). The main classified areas and all critical stability chambers, and critical chemistry and microbiology storage systems must be backed up with a power generator (either natural gas or diesel); these systems must be very reliable and automated such that they are always on standby and ready to start in case of power failures. Putting the aseptic building on different power grid supplies from the electric company must prove to be useful with allowances for automatic switching from one grid to another. In all cases, the impact of power interruptions must be properly evaluated and documented as it might affect the sterility of the products.

The advancement in lighting moving towards LED lighting provide for more luminacins and should be measured at least 500 lux 3 ft from the floor. Amber lights might be used in ISO 5 suites to protect the product from UV light interference with product quality. Lighting heat generation must be incorporated in the design of clean rooms to control product temperature. All light fixtures must be sealed and cleanable to be free from particulate accumulation

Materials of construction of light fixture must be compatible with classified room requirements. Special provision for inspection lighting and backgrounds are usually specified for such applications. Automated visual inspection must have neutral lighting that will not interfere with camera function during inspection. All wiring, outlets and associated equipment must meet the US electrical code as well as the permits for the city in which equipment and facilities are being prepared for manufacturing operations.

17 Sterile Techniques

Sterile Filtration is used when solutions are not sterilizable through lethal heat treatment. This requires the solution to pass through a 0.22-micron sterile filter. The solution might be heated to a certain limit without impacting compounded ingredients including API efficacy. Sterile filtration intent is to reduce micro loading to render solutions sterile. As sterile filtration has limited effects on endotoxins, the upstream solution must be maintained in relatively clean transfers with lower bioburden to minimize the formation of endotoxins. Closed systems are usually sterilized through SIP and can be cleaned through CIP processes.

Sterile filters must be validated for materials compatibility with the solution to prevent leachables and extractables. In addition, filter integrity must be tested before and after sterile filtration is applied. The sterile filter must not alter the solution concentration by retention of some of the formulated ingredients in the compound. Similarly, it should not release any of the chemical composition of the filter structure into the solution. Filter wetting and surface tension characteristics must be validated for micro retention. Overall, filtration system with all associated piping and connections must be pressure tested using an inert gas. All drain line connections to sterilizing piping must be fitted with valves and air breaks to prevent backflow into the system leading to contamination.

During the component preparation processes on washing sterilization or depyrogenation the bioburden of viable micro counts is reduced to sterile levels. Endotoxin control of pyrogenic bacterial cell wall materials resulting from growth and degradation of microorganism must be monitored along the process.

During handling of components, some breakage might occur and creates extraneous particulates that might transfer into the product. In addition, residual chemicals used during washing must be checked through the validation of cleaning cycles.

Vials might be washed in ISO class 7 using automated washers. The automated washer can be connected to a depyrogenation tunnel that would transfer the vials directly into the ISO 5 filling line, which will be followed by capping and tray-off. Partial stoppering and loading into Lyophilizers is also accomplished in a similar manner. Following the completion of the Lyo-cycle, vials are stoppered and transferred through unloader to capping and tray-off. WFI rinse is used before depyrogenation to control endotoxins. Depyrogenation tunnels are fitted with unidirectional flow to control processed glass temperature in the various zones of the tunnel.

Filling systems are advanced to control volumes up to 0.1 ml. Liquids are easier to control and fill than powders. Liquids are metered and filled by weight overload cells. Solutions are transferred through piston or peristaltic pumps using flexible tubing. The selection and control of the tubing movements during pumping is critical to aseptic conditions. Filling temperatures and recirculation in aseptic tubing is critical to the maintainability of condensation on equipment and containers. Weight control of filled volumes is critical in filing cartridges and pre-filled syringes for dosing purposes. Vacuum and sterile inert gas purge might be applied before final sealing of containers. This is important for oxygen-sensitive products that might interfere with product stability. Powders are handled in vacuum transfer and predetermined volumetric canisters at the filling station. Gravimetric systems are also utilized to control filled weights, which are sensitive to powder density and moisture content, and the flowability of a particular compound.

Sterile filling lines are built with ISO 5 RABS or isolator to minimize contact with product, containers, and closures in aseptic manufacturing. All product contact surfaces should be SS 316L. Depending on the type of operation all surface contact must be easily cleanable. In fully automated processes, provisions for CIP and SIP or VHP should be made to allow for sterile environment preparation inside the machine and the filling line stations. Gloves access ports in critical positions are important to minimize operator interventions. All recirculation air must go through HEPA filtration. Stopper bowls are usually a challenge in terms of cleaning and sterilization. If bowls cannot be sterilized in place, then they must be removable for autoclaving purposes.

BFS system utilizes extruded high-grade virgin plastic into a parison (hollow tube), which is then presented to a specific container mold. As the container is shaped in the mold, the product is filled into the container and sealed before the product is released from the mold. Filled containers are checked for filled weight and integrity.

Lyophilization is a process that is reserved for products that need to be stabilized as a powder cake in a vial. In addition, API and excipients are sensitive to heat and their efficacy might be lost due to lethal heat sterilization processes. The process starts with a clean chamber, followed by a vacuum integrity test to ensure that the chamber is airtight. The chamber is SIP sterilized. Partially stoppered liquid solution filled vials are introduced through the main door in manual operations, but mostly if using automated loading then product is introduced through a slot door and stacked on shelves that are engineered apart for positioning. Following loading, freezing of the product in the chamber (-40 C) is accomplished by running silicon oil through the tubes built in the shelves or through direct expansion fluid. Following freezing, an interface Valve is opened, and vacuum is applied to create primary drying sublimation. After sublimation, secondary drying desorption brings the product to a specific moisture content based on cycle development requirements. The chamber is back filled with inert gas and stoppering occurs by bringing the shelves down over the vials. Following equalization of pressure through aeration unloading of the product can be accomplished manually by opening the main door, or more through automated unloading systems. The condenser attached to the chamber is then defrosted (-70 C), followed by CIP. Two nitrogen filters in series are tested for filter integrity, followed by a leak test. All these processes are PLC controlled and are driven by step-by-step software that is designed for various recipes. All software recipes are validated under a CSV program with GAMP 5 active directory and multi-level password access requirements. Lyophilization is a time-consuming process, which takes about 72 hours to complete a batch. Multi Lyophilizers are installed in parallel to match production capacity of a filling line. Cooling/freezing systems can be based on multi-stage compressors or liquid nitrogen freezing system.

Products that incorporate alcohol of more than 22% in the formulation, usually require special vacuum design and a controlled release environment to mitigate the risk of explosion and prevent fire. Vapor release and exposure limits must be included in the design provisions.

Capping and crimping are introduced to secure the stoppers or septums in place to assure the seal integrity of the vial or cartridge. Following cap placement, a crimp is applied to tighten the seal. These processes might generate metallic particles and should be separated physically and through pressure vacuum systems to prevent particulate matter from entering the stoppering area.

Component sterilization utilized moist heat–saturated steam. Filled product containers are sterilized using superheated water or through a steam/air mix. The equalization of product inside the sealed container due to internal pressurization during sterilization needs the provision of air or rain on product with super-heated water to prevent bursting of containers. Forced air using fans might be employed to ensure equalized heat penetration.

Visual inspection might be manual, semi-automated (Seidenader), where the product is conveyed and rotated in front of an inspector, or fully automated using cameras, whereby pictures are analyzed for defects using computer modelling. The entry and discharge of containers into the visual inspection machine requires automation for handling of product containers and unloading from feed conveyors and loading into shipping containers upon discharge from inspection (Brevetti / Marchesini). Advanced systems can emulate human defect selection that is confirmed through Knapp studies. In addition, the reject stations are selective by defect type and record errors accordingly for statistical process control.

Products contained in amber glass, emulsions and suspensions, Lyo powder cake present challenges that must be extensively tested during validation. The use of Instron equipment testing for cap seal integrity deflections and other leak testing methods using customized fixtures are important features that are developed as part of companies' core competencies. Moisture can be measured in terms of bound water or free water. Headspace analysis can be measured to determine oxygen levels in sealed containers after sealing using inert gas blanketing or vacuum /gas flush systems.

Online label printing has been recently improved by adding serialization systems for individualized product units and brick type units used as shipping module for traceability. Batch traceability along the supply chain is now an important regulated aspect of product packaging. Label reconciliation to reduce risk of mix-up is an important step that needs verification of correct labeling taking into considerations reliability of printing system and failure potential.

Bacterial contamination and bioburden reduction are key factors to consider in the design of aseptic facilities. The control of mold is as important when dealing with wet surfaces. The use of fumigation in classified clean space utilizing VPHP has proven effective if the fumigation automizer is properly designed and the fumigant reaches all contact surfaces. Similarly, the use of Minncare sterilant to chemically sanitize RO systems might be effective to ensure cleanliness of purified water systems. This application can be performed on a biweekly basis. Sanitization using hot water at 80 C is also used in distribution loop systems and CIP skids. Testing of 20–30 min time cycles must be confirmed depending on the design of each distribution system or skid. Confirmation of bioburden reduction in laboratory testing is required.

Fire protection in aseptic facilities requires all sprinkler systems to be enclosed in classified areas and all sprinkler heads to have a pop-up design that will protrude from the ceiling by pushing the speciality covers. In addition, all in-line baffles installed in the HVAC delivery system must fail closed in case of fire to prevent and isolate fires in the local suite

Materials of construction in cleanrooms are of specialized types. From the full ceiling coverage by HEPA filters in ISO 5 suites, the walls might be covered with 0.08" thick vinyl sheets that are welded together. This type of wall provides for wet cleaning and moping after each batch. Specialty epoxy paint can be used to create a non-shedding, non-porous and resistant to microbial growth type surface. The inside wall materials should be free from kraft cellulose type materials that would be feeding ground to molds and other microorganisms. Floors are usually covered with epoxy or polyurethane polymers that are not porous and the coving continues to about 4" in height at the point of meeting the walls. There should be no seams or breaks in the materials covering walls and floors in ISO 5 rooms. All surfaces must be inclined to prevent deposits on ledges or windowsills. The whole structure must be easily cleaned and have no hidden corners for particulate matter to accumulate. All surfaces must withstand daily repeated cleaning with sanitizers and disinfecting chemicals. There should be no materials that are susceptible to oxidation *and* corrosion. Doors must be sealed and have no access to cleaning solutions to stay inside the door. Experience shows that such events will be host ground for molds. All doors must be latched properly to provide for design differential pressures. All doors must be interlocked to move from one room to another to minimize air movement and potential decontamination. All door closures must be made of sanitary type materials and should be fully enclosed.

HVAC systems are critical utilities that support clean rooms air circulation. ISPE baseline provides for extensive detail related to the design and installation of AHU that provide HEPA filtered air to classified zones. Following commissioning, the as built conditions of an HVAC system must be balanced among the aseptic core and adjacent cascading rooms to ensure that pressure differentials are met and positive pressure exists from the aseptic core all the way out to wrap around CNC space. Environmental data monitoring for the aseptic area should match the in-operation data collected during equipment and room qualifications. Critical product and process parameters are greatly influenced by room operation conditions, specifically temperature and humidity. Temperature might affect liquid products filling operation flow characteristics and fill volumes. Humidity will affect moisture content in powder aseptic filling. A recommended temperature range in aseptic filling is 62–68 F, which provides for comfort level to the heavy gowning in ISO 5 rooms. Lower humidity levels might create static electricity. Higher humidity might cause condensation on filled product containers. Humidity conditions affect labeling, which requires appropriate relative humidity levels for label application. GEP guide in keeping humidity levels below 60%RH.

One critical area in ensuring continuous air flow through the HEPA filters is to constantly monitor air flow from each HEPA filter by taking measurements routinely at the face of each HEPA filter, whether the installation is directly air supplied HEPA or through a common planum. The minimum reading in ISO 5 should be 72 ft/min (nominal 90 ft/min) at workspace under unidirectional laminar flow conditions. It is important to design and build a robust AHU system to insure uninterrupted operation. Nonetheless, the HVAC system design should be able to recover within 15–45 min as a measure of overall system effectiveness.

Establishing differential pressure between cascading rooms is necessary to direct air flow from the higher-pressure aseptic core to the adjacent room space. A minimum recommended DP of 0.002 psi (0.05"-water) should be maintained between adjacent rooms or between RABS and surrounding space.

It is important to build a simple system that can be easily balanced between the rooms as cascading occurs. Airlocks play an important role in the separation of classified areas. Airlocks have limited volume and are designed for quick recovery after doors are closed.

Monitoring DP is critical to Classified rooms operation. This can be accomplished by installing pressure sensors in classified areas (One sensor per room) and connect all readings through a pressure controller to the PLC and SCADA on the BMS. Irrespective of loss in pressure, DP should never be zero or reverse from one room to another. Therefore, it is important to maintain airflow and ACH per room as a constant. HEPA filters should also be checked for particulate loading that might influence air flow conditions. Filters must be placed on periodic check for air velocities at the face of each filter and should be on a replacement schedule as needed.

18 Compliance

Compliance strategies require thoughtful coordination and guidance with respect to projects harmonization, standardization, and integration. Under the major umbrella of process validation/equipment qualification a work plan must be developed to deliver a documentation package to cover all areas under remediation, namely: WFI, STS, HVAC, Mixing and compounding classified areas, and clean filling suites with emphasis on systems design and installation qualifications including all controls managed by all requirements governed by computer systems validation (CSV). A similar approach for Chilled Water (CW) system upgrades could be followed.

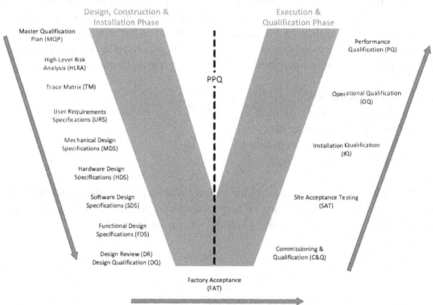

A V-model for each Key (Master) project and sub-projects would be recommended as a road map to achieve full compliance:

I – DESIGN, CONSTRUCTION, AND INSTALLATION PHASE

- Master Qualification Plan (MQP)
- High-Level Risk Analysis (HLRA)
- User Requirements Specifications (URS)
- Traceability Matrix ™ (TM)
- Mechanical Design Specifications (MDS)
- Hardware Design Specifications (HDS)
- Software Design Specifications (SDS)

- Functional design Specifications (FDS) – at this stage, will highlight the linkages with URS through the traceability matrix
- Factory Acceptance Testing (FAT) – As needed
- Design Review (DR)/Design Qualification (DQ)

II – EXECUTION AND QUALIFICATION PHASE

- Commissioning & Qualification (C&Q)
- Site Acceptance Testing (SAT) – As needed – At this stage following SAT, there are commercial indications such as the start of warranties on Labor & Materials including equipment and instrumentation.
- IQ/OQ/PQ

Specific attention to HVAC systems for classified rooms qualifications, WFI, STS and all critical systems shall be documented in detail designs. Protocols and reports to document all qualification executions including all required micro and chemical testing must be written and executed. SOPs must be generated or updated as needed.

To ensure continuous compliance, the following highlights are of importance:

- Mix-up prevention
- WFI trending data (Micro & Chemistry)
- Pest Control Trending data
- DOE-based cleaning validation
- Endotoxin LAL trending

HEPA filters should also be tested for leakage due to tears in the structure of the filter. An upstream concentration of aerosol that will produce an indication downstream if leakage exists (See ISPE guide-HVAC for more details.)

19 Controls

User Requirements Specification (URS) for Automation System applies and considers the normal range of System operation. User requirements specify equipment that are to be used to support the production of finished pharmaceuticals in a cGMP environment.

URS is an integral part of a procurement agreement with the selected equipment vendor. Equipment supplier or vendor abides by the information and conditions set forth by URS.

An equipment validation plan is developed to outline the planned tasks and expectations for qualification of the equipment. URS serves as the input to subsequent project risk assessment activities, design objectives, control strategy, and acceptance criteria for testing and qualification of the manufacturing system. The requirements described will outline the general and basic needs of the system.

Control systems are defined as direct impact systems based on their anticipated use. URS describes at a high level the functions that control automation is required to perform each requirement will be uniquely identified.

The Automation System interfaces with other process systems including but not limited to Solution Transfer System, Water for Injection System and Powder Transfer System. Control system provides for total control of the batch process.

Batches are started from the MES system, and electronic batch record is usually stored in the MES system. Automation System receives batch parameters from MES and provides batch data to MES. All data interface is handled through the FactoryTalk (FT) Batch server.

Current industry, association, and regulatory standards and guidelines are followed as applicable to the design, fabrication, and assembly of all equipment furnished under URS.

ANSI	American National Standards Institute
ASME	American Society of Mechanical Engineers
ASTM	American Society for Testing and Materials
AWS	American Welding Society
cGMP	Current Good Manufacturing Practice
EN	European standards
FDA	Food and Drug Administration
ICH	International Conference on Harmonization
IEC	International Electrotechnical Commission
ISPE	Int'l Society of Pharmaceutical Engineers
NEC	National Electrical Code
NEMA	Nat'l Electrical Manufacturer's Assoc.
NFPA	National Fire Protection Association
OSHA	Occupational Safety and Health Act
UL	Underwriter's Laboratories
21CFR	

Term	Definition
Historian	A server-based application that pulls data from remote systems for archival and reporting.
Installation Qualification	Verifying and documenting that the equipment exists and is installed per specification.
Operational Qualification	Verifying and documenting that the equipment can operate within established criteria.
Performance Qualification	Verifying and documenting that the equipment/system reliability and repeatability produces the desired result for the intended processes.
Server	A network-based computer which allows access from remote systems for application or data retrieval.
Specification	A list of tests, references to analytical procedures, and appropriate acceptance criteria that are numerical limits, ranges, or other criteria for the test described. It establishes the set of criteria to which a material should conform to be considered acceptable for its intended use.
Thick Client	A computer that is designed to run stand-alone or in a networked environment. Often referred to as an Engineering Workstation for application development.
Thin Client	A computer that is designed to run from a server-based application with minimal hardware specifications

Abbreviation	Definition
EMS	Environmental Monitoring System
CE	Certificate of European Safety
CIP	Clean in Place
CPU	Central Processing Unit
FT	FactoryTalk (Suite of services developed by Rockwell Software)
HMI	Human Machine Interface
ISO	International Organization for Standardization
MES	Manufacturing Execution System (EBR)
NEC	National Electric Code
NEMA	National Electrical Manufacturers Association
OIT	Operator Interface Terminal
PAC	Programmable Automation Controller
PID	Proportion/Integral/Derivative – Algorithm for Process Control Loops
PLC	Programmable Logic Controller
SCADA	Supervisory Control and Data Acquisition
SAT	Site Acceptance Testing

Automation system is implemented in support of complex mixing operations. A turnkey mixing control system is integrated with mixing equipment and MES system. The MES system directs the starting of a batch. Automation system is designed to control equipment and monitor instrumentation that produces a batch. Electronic Batch record is maintained by the MES system. Controls include an automated powder transfer system (PTS), Solution compounding and mixing, STS, Temperature Control Unit (TCU) for CIP system(s), and addition of one or more excipients. The Control system will interface with WFI system for water addition to the mix tank, and the STS system for transfer of the batch to the fill lines. The Control automation system provides for consistency between batches and reduces the likelihood of an operator error resulting in a batch not being made according the the product's batch record.

Mixing complex is comprised of multi mix tanks that are operated in sets to serve filling lines. Each line will utilize a set of tanks. Controls will operate tanks independently, e.g. filling from one tank while mixing or cleaning the other. Control automation systems operate using preconfigured "recipes". Based on the materials required for any given solution, the system will require inputs from various sources.

LIFECYCLE REQUIREMENTS

Equipment Suppliers are expected to provide information during the manufacturing cycle of a project:

- Written communication of any deviations from specifications.
- Single point of contact for communication.
- Provide an SDS, HDS, and FDS.
- Provide FAT, SAT documentation
- Participate in FEMA discussions.
- Written communication of changes to the design specification.
- Manufacturing/completion project schedule.
- Written communication of delays in the schedule.

Process steps are as follows (beginning with clean equipment). There may be confirmations by a second operator for certain steps. This is only a high-level description meant to outline the process.

1. Batch information is entered in the MES system prior to the start of batch processing.
2. Operator selects the appropriate batch and starts the process from the MES system.
3. MES system will send batch-specific information such as product code, batch number, and batch size to the Controls system.
4. Operator will need to interface with both the Controls system and the MES system processes. The operator will be prompted by both systems to intervene as needed.
5. Controls system will control and monitor instrumentation. If the process is stopped unexpectedly due to a power loss or emergency stop, the process will be paused in its current state.
6. Batch will stay active in the control system until the filling operation is completed.

7. Once the batch is complete, an operator in Filling will end the batch and release the unit to allow the mixer to move to the cleaning stage.

8. When cleaning is required, the CIP system engages to perform cleaning. The next batch cannot be started until the cleaning process is complete.

PRODUCT QUALITY REQUIREMENTS

Critical Quality Attributes (CQAs)

Process Step Reference	CQA	Specification	Acceptance
Initial water addition [1]	Initial Weight	+/- X% of weight	Load cell reading
QS	Final volume	+/- X% of weight	Load cell reading
Agitation	Agitator speed	+/- X% of set point	Hz measurement
Agitation	Mixing duration	+/- X seconds	Time measurement
Temperature control [2]	Temperature	+/- X° F of set point	Temperature measurement

[1] Temperature limits required for some product's initial WFI addition.
[2] Applicable only to jacketed tanks.

PROCESS PARAMETER REQUIREMENTS

Critical Process Parameters (CPPs)

Process Step Reference	CPP	Operating Range	Associated CQA
Initial water addition	Weight	within X% of weight	Initial volume
QS	Weight	within X% of weight	Final volume
Agitation	Drive speed	+/- 5 Hz	Agitator speed
Agitation	Time	+/- X seconds	Mixing duration
Temperature control [1]	Jacket water temp	+/- X° F of set point	Temperature

SYSTEM BOUNDARIES AND REQUIREMENTS

The following list represents the boundaries that the system will be capable of:

The system will be configured to run different recipes.

Recipes will include automated and manual additions.

Specified tanks will have temperature requirements.

Specified products have WFI temperature requirements that the controls system needs to monitor.

Specified products require a pre-mixed slurry to be added from an auxiliary tank.

Specified products require additives to be added manually

Specified products require bulk fed powders from an automated Powder Transfer System.

Requirement Key:

M	Mandatory	This requirement must be met. It impacts critical parameters; product quality, safety, or efficacy or personnel safety. The equipment/system is not acceptable unless this requirement is provided in an agreed manner.
I	Important	The equipment/system will provide significantly greater, product quality, safety, or efficacy, personnel safety or business benefit if this requirement is met but, at the Owner's discretion may not make the system unacceptable if it is not delivered.
D	Desirable	These requirements might be worthwhile, but it will not make the system unacceptable if not delivered.
IN	Information	Information provided for the supplier which may be useful for equipment design or calculation of the offer.

Automation

	Requirement			
Requirement Statement	M	I	D	IN
URS for Overall Automation requirements.	☒	☐	☐	☐
One dedicated Ethernet Module shall be used for interconnection to the Enterprise data servers and the FT Historian.	☒	☐	☐	☐
A separate Ethernet Module shall communicate to a dedicated Ethernet network to provide interconnection between the control processor and its associated remote-mounted I/O using a fault-tolerant configuration.	☒	☐	☐	☐
The loss of power to any control remote I/O cabinet shall not cause loss of Ethernet communication to any other control remote I/O cabinet.	☒	☐	☐	☐
Controls will be capable of having multiple thin clients that can be accessed simultaneously.	☒	☐	☐	☐
Electrical systems must be designed per the Electrical-Controls Equipment Specification provided	☒	☐	☐	☐
All Electrical panels should have externally mounted disconnects for Arc Flash compliance.	☒	☐	☐	☐
Panel, field wiring, devices, and instruments must be permanently labeled to match the drawing package at all termination points.	☒	☐	☐	☐
Power supply feeds must be set to 480VAC 3 Phase or 120/240 Single Phase	☒	☐	☐	☐

(*Continued*)

Requirement Statement	Requirement			
	M	I	D	IN
Manual mode will be used during maintenance and troubleshooting to control individual devices	☒	☐	☐	☐
The system will include the application software, Server Unlimited Graphics, FT Batch Unit, FT Historian 2500 Tags, SQL Server	☒	☐	☐	☐
Each fill line will have a main control panel. The main control panel will include; Control Logix processor, multiple ethernet cards, Managed ethernet switch, 24 VDC power supply and required I/O.	☒	☐	☐	☐
Each fill line will have a valve manifold panel which will include; disconnect, managed ethernet switch, process solenoid manifold for up to 64 valves and feedback	☒	☐	☐	☐
The CIP system will have a main control panel. The main control panel will include; Control Logix processor, multiple ethernet cards, Managed ethernet switch, and 24 VDC power supply	☐	☒	☐	☐
The CIP system will share the valve manifold with each fill line. The valves will be controlled through produced and consumed tags.	☐	☒	☐	☐
The CIP skids will have a remote I/O panel, and the VFD drives required to control the skids. These controls will be provided by the skid vendor	☐	☒	☐	☐
One desktop thin client HMI station will be provided	☐	☒	☐	☐
Multi wall mount thin-client HMI stations will be provided. At the PTS level At top of Tank level At bottom Tank level	☐	☒	☐	☐
The Controls, CIP, STS and WFI HMI applications will be set up on EBR work stations throughout the Mix Complex	☐	☒	☐	☐
EBR work stations will interface with HMI setup	☐	☒	☐	☐
Training and work instructions will be provided for setting up the HMI applications on the EBR workstations, laptops, and tablets.	☐	☒	☐	☐

Equipment Alarms and Warnings

Requirement Statement	Requirement			
	M	I	D	IN
URS for Overall Equipment Alarms and Warning requirements	☒	☐	☐	☐
Discrete device valves, actuator position must be monitored in both directions: open and closed	☒	☐	☐	☐
Any CQA that is out of specification limits must trigger an alarm	☒	☐	☐	☐
System compressed air pressure outside of limits must trigger an alarm	☒	☐	☐	☐
Mix tank fill rate shall be monitored during fluid addition. A fill rate lower than the set point must trigger an alarm	☒	☐	☐	☐
After powder addition by PTS tank weight load cell shall be used to verify most of the material has been added to the tank.	☐	☒	☐	☐

Safety Requirements

		Requirement		
Requirement Statement	M	I	D	IN
There will be two safety zones per line. One at the top of tank one at the bottom. The safety zone at the bottom of the tank extends to the fillers	☐	☒	☐	☐
The design must permit the electrical, pneumatic, and hydraulic devices to be locked out during maintenance and repair activities.	☒	☐	☐	☐
The control system must be designed to operate in a fail-safe manner and meet OSHA safety requirements.	☒	☐	☐	☐
Safety circuits must not be able to be bypassed.	☒	☐	☐	☐
Fixed mount HMIs are to be set at a height of 54" from the floor to the center of the screen. Adjustable mounts are to cover the range of 54"–68" to center of the screen.	☒	☐	☐	☐
An emergency stop button will be located at the top of the mix tank in each mixing room. The emergency stop will stop water addition, tank jacket heating, and CIP flow.	☒	☐	☐	☐
An emergency stop button will be located at the bottom of the mix tank in each mixing room. The emergency stop will close the bottom tank valve, Stop STS, and CIP pumps	☒	☐	☐	☐
Two "emergency" stop buttons will be in each fill room. One near each filler. The emergency stop will close the fill supply valves	☒	☐	☐	☐

Process Control System

		Requirement		
Requirement Statement	M	I	D	IN
Control system will allow automated and manual component additions.	☒	☐	☐	☐
System will allow a 2nd person to verify critical manufacturing steps	☒	☐	☐	☐
System will receive and supply data to MES	☐	☒	☐	☐
System will utilize distributed I/O panels strategically located throughout the complex.	☒	☐	☐	☐
Process interlocks will be implemented at the control module level and should prevent undesirable controlled process situations.	☒	☐	☐	☐
During the batch process when a critical alarm occurs, the batch will be put on hold. Depending on the nature of the alarm the batch may be continued or aborted.	☒	☐	☐	☐
At any time during the batch process the batch may be on hold or aborted by a user with the appropriate privileges.	☒	☐	☐	☐

Instrumentation

Requirement Statement	M	I	D	IN
Weight Transmitter	☒	☐	☐	☐
Up to four temperature probes (4 per tank).	☒	☐	☐	☐
Agitator VFD Drive	☒	☐	☐	☐
Pressure Sensors	☒	☐	☐	☐
Conductivity Transmitter (CIP)	☒	☐	☐	☐
Flow Transmitter (CIP)	☒	☐	☐	☐
Level Transmitter (CIP)	☒	☐	☐	☐
Flow Switch (CIP)	☒	☐	☐	☐
Temperature Control Valve (CIP)	☒	☐	☐	☐

Training and Documentation Requirements (DOC)

Requirement Statement	M	I	D	IN
Training documents will be supplied for Operation procedures.	☐	☒	☐	☐
Operator training will be available during commissioning, potentially on multiple shifts.	☐	☒	☐	☐
Maintenance-specific training will be available during commissioning, potentially on multiple shifts.	☐	☒	☐	☐
A draft Operations Manual will be available during the FAT and the final version at the IQ. The operations manual will include HMI screenshots with description of screen object, special instructions, electrical spare parts list, etc.	☐	☒	☐	☐
Documentation will be provided in both electronic and hard-copy versions	☐	☒	☐	☐
Electrical system required documentation must include: • Electronic copies of all equipment programs or parameters • Electrical Prints Package • Recommended Spare Parts and BOM	☒	☐	☐	☐
OEM part numbers are required for all spare parts.	☒	☐	☐	☐
Spare parts lists will be available one month prior to FAT.	☐	☒	☐	☐
AutoCAD drawings will be supplied for all electrical panel layouts	☒	☐	☐	☐

CRITICAL PARAMETER CONTROLS

Classified rooms temperature is always actively controlled and measured with continuous recording. RH should always be actively monitored with continuous recording. Continuous control and recording of differential pressure are required. Continuous particle counts using CPMS is recommended and should be controlled with alarms for alert and action levels. Alert alarm indicates that a critical parameter deviated from normal operating range but still within limits. Action alarm indicates that a critical has deviated from process limits, which will require immediate attention as it might have an impact on product quality. A deviation is usually issued to assess and justify an action alarm.

The use of instruments and sensors is more in use as advances in controls are more prevalent. Performance of these controlling elements is totally dependent on accuracy, resolution, repeatability, hysteresis, response time, and stability. Fit form and function are important in the selection of instruments and sensors. Deviations in these elements are based on ranges and limits to their capabilities but must not impact the safety and quality of drug products. Higher accuracy instruments reduce the risk of manufacturing under wider range of conditions due to instrument drifts during processing. Critical process parameters are driven by boundary layers or process limits within which a product must be produced. These limits must be documented in product specifications and process validation. Control systems instrumentation are designed with set points based on proportional or integral conditions with margins of safety during operational conditions. Temperature, flow rate, and pressure readings are important process parameters and variables that would affect the outcome of product quality. Concentrations of fluids and residual impurities will greatly affect the purity of API and drug products formulations. Instrument readings have built-in tolerance, which should include drift allowances. All instruments and sensors must be calibrated and continuously monitored for accuracy. All instruments and sensors must be logged into a tracking system to ensure that periodic calibration (3–6 months) is routinely conducted with special attention and shorter time spans between calibrations for critical measurements.

Computerized systems lifecycle activities should be scaled up and based on design and verification risk-based system impact on patient safety, product quality, and data integrity. System architecture and components must be evaluated for complexity and categorized accordingly.

ELECTRONIC BATCH RECORDS (EBR)

Electronic batch records are established through the manufacturing execution systems (MES) to control the manufacturing process and interact with plant control systems that manage the process parameter controls for manufacturing operations. EBR provides for batch scheduling (amount of material and WFI for each compounding (mix) tank). It also covers materials management (Inventory, Checks, Weigh Minor Ingredients). EBR manages area approbations, tank availability checks, labeling (material, and tank status), cleaning of individual bulk container (IBC), slurry tanks, slurry mix, pH additions, and adjustments. It also manages mix tank sampling and adjustments for all mix tank release. In support of EBR requirements, plant control systems provide for tank WFI addition, tank agitator control, tank bulk material addition from powder transfer system, tank temperature control and data storage, compounding recipes (sequence of phases required to mix a tank), distribution of solution to fillers (Solution Transfer System), cleaning of mix tanks and STS (Including CIP), and reporting mix process information to EBR.

Mixing Process Flow

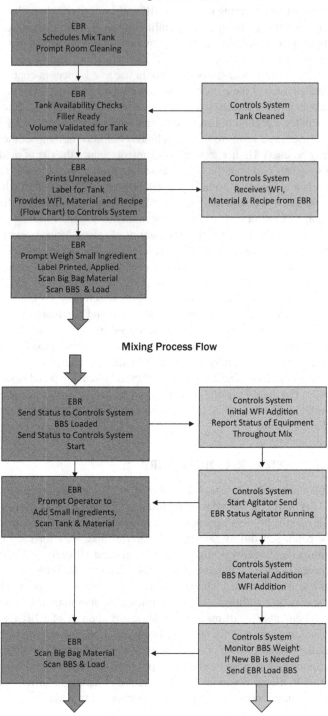

Mixing Process Flow

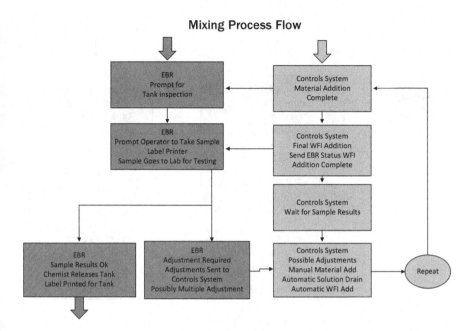

Mixing Process Flow

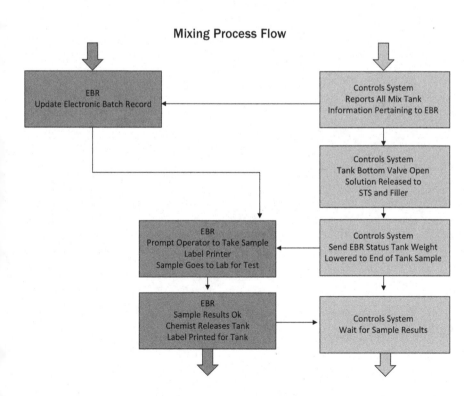

Mixing Process Flow

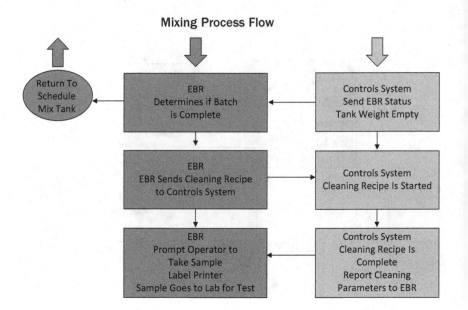

20 Barriers and Isolators

Restricted Access Barrier Systems (RABS) and Isolators are now important considerations in the design and construction of filling and capping lines. These systems can be fully or partially automated to include interlocks, alarms, and complete processing including CIP and SIP (VHP). Mainly, these setups are intended to reduce and mitigate the risk of operator interventions, which might impact product integrity during aseptic filling. These systems can be small-scale for filling clinical trial formulations, or large-scale to entertain commercial high-speed automated lines.

Isolators might be placed in an ISO 8 zone. RABS are built around filling lines in ISO 6 zone with target ISO 5 aseptic core inside the barrier. Design considerations for optimized operations is critical to the use of these systems. Accessibility, intervention, cleaning, sanitization, and sterilization are all process requirements that are used to ensure functionality of these barriers. Unidirectional flow and pressure inside these systems is very controlled to ensure that particulate matter is as specified.

Active RABS use an integrated HEPA filtered air supply over the filling and capping stations. Cleaning and disinfection are introduced through the use of active agents such as VHP. Gloves and transfer path-throughs are used for materials movements with the aid of specialty tools.

Passive RABS air flow is provided by room ceiling HEPA filters. Unidirectional air flow is directed over the filling and capping stations inside the RABS and balance to flow inside and outside barrier walls to control particulate from concentrating in the surrounding environment. Gloves and gauntlets are used and controlled by SOP as they are attached and removed from glove ports in the critical locations on the RABS or Isolator systems. Gloves are sterilized and disinfected or changed as established by validation protocols.

RABS and Isolator systems are superior to just cleanrooms. They require a higher ceiling and accessible foot print around these self-contained units. Isolators can be built with complex installations on the inside and thereby require extensive validation protocol to ensure validity of repeatable operational procedures. There are tradeoffs for operating cost and manufacturing hurdles using these systems. Environmental reproducible decontamination with automated cycles should be validated, but as operators are removed from the direct filling/processing environment, products are more protected and operator safety is enhanced with the use of potent compounds. Once line operation is established, cleaning and handling complexities become routine. Nonetheless, as line manufacturing step are assembled in side an isolator, systems integration and control can be complex as transfer occurs from one station to another inside the isolator. Line changes involving size, change parts, and component change will require product change cleaning and line clearance, which might not be as straightforward, but becomes easier as procedural methods are applied.

When products are manufactured inside an isolator, the gowning level of operating personnel would be eased to a lower classification level, which increases operator comfort under strict temperature conditions. Access to aseptic processing areas where isolators are put in operation does not require sterile gowning, but more like gloves, goggles, hair net, beard covers, scrubs, and shoe covers. RABS do not offer relaxation of sterile gowning procedures.

Dust and dust explosions are of concern in the use of powder transfer systems. Isolators and other containment methods will mitigate these effects as the isolator will shield operators from the processing environment. KST values must be calculated as part of the design while handling powders that might be susceptible to explode depending on oily contends, fine milling particle size, surface tension, angle of repose and other ignition factors that must be vetted to prevent catastrophic failures. However, isolators can be useful for processing hazardous powders or biological potent materials when operated inside the closed system. Toxic compounds can also be handled inside an isolator but are limited while processing in RABS.

ISOLATOR DESIGN CONSIDERATIONS

As with a confined space heat can be generated due to space limitation. Heat loads inside an isolator should be carefully calculated to manage temperature control. Besides ergonomics concerns the positioning of glove ports is critical to specific process transfer of containers inside the isolator if operator interface interactions are required including any sampling. The use of sterilant and generated vapors and effects on both materials of construction and operators should be well defined. Sterilant such as chlorine dioxide or hydrogen peroxide, peracetic acid are all aggressive oxidizers and their use must be documented step-wise in SOP. Rapid transfer ports should be built with flexible connectors to ensure that the aseptic core remains intact. Laminar flow inside an isolator needs to be controlled for recirculation through built-in HEPA filtration system. If powders are handled inside the isolator, special design consideration must plot the flow regime and the accumulation of powder on the various surfaces. All considerations for dust control and decontamination for batch start-up requirements are crucial to clean operations. Positive pressure inside the isolator prevents ingress of any contamination from the surrounding environment. Nonetheless, balanced design pressure should consider specific application when dealing with high potency drugs that must not exit the isolator into the surrounding environment.

21 Quality Risk Management

ICH Q9 describes the technical requirements for pharmaceuticals' human use based on a systematic approach to quality risk management. The fundamental principles of risk management emphasize scientific methods, experimentation, and testing links to efficacy and safety of products for the protection of patients. Levels of risk are clearly spelled out as harm causing damage to health, hazard source of harm, risk probability of occurrence, and severity measure of hazard consequences. Assessing risk through identification and analysis of event evaluation, control, mitigation, and acceptance of results are process steps that should be clearly reviewed and communicated to concerned bodies.

Finding the problem is significant in terms of identification and characterization before any attempts to present solutions. Evaluating potential risks to build in prevention as part of an FMEA to safeguard process interventions and out of specification products is at the core of risk mitigation. Therefore, data gathering, and assembly of background information, is essential and relevant to risk assessment. Risk assessment should be managed through formalized project leadership with appropriate resources, a timeline and specific tasks should be drafted as part of a risk assessment schedule. Impact analysis to evaluate operational tolerances, facilities and equipment fit, and functionality are all important factors in this assessment.

Following risk assessment, risk control and reduction should be implemented to address risk -cceptable levels. Questions to be answered would be: What can be done to reduce or eliminate risk in specific event situation? What would be an appropriate balance between benefits, impact, and available resources and resource allocations? Besides preventive activities such as the FMECA, as situations arise, lessons learned to prevent recurring risk should be based on root cause analysis team participation.

22 Equipment Qualification (EQ)

These guidelines are to establish a uniform harmonized approach to equipment qualification and re-qualification of all equipment used for the manufacture of drug products, medical products, or combination of both fill/finish delivery product systems to assure compliance with regulatory requirements.

In general, qualification should begin with the planning of new equipment or changes to existing equipment. That is, an impact assessment will be made when a change is being considered to determine the extent of qualification required.

Critical systems and equipment will be qualified if they are in direct physical contact with the product or if they are used to measure, monitor, or record a critical parameter.

Indirect or no-impact systems and equipment such as chilled water, plant steam, electric power, and non-process water systems do not require formal qualification. They will be qualified and documented according to good engineering practices only.

The minimum requirements for each system to be qualified are a qualification plan and adequate documentation to demonstrate that all pre-determined acceptance criteria were met.

Prospective EQ consists of four distinct qualification phases: DQ, IQ, OQ, and PQ. Changes to qualified equipment do not necessarily require all four phases of qualification, depending on the extent of the change being considered.

Qualification by suppliers and contractors is acceptable if they have been qualified by the Quality department and approved in advance. Additionally, their work must be reviewed, initialed, and dated by the responsible Quality representative.

Only critical Chemical Development equipment used for the manufacture of chemical intermediates intended for use as clinical trial material requires qualification. Laboratory scale manufacturing equipment in chemical development does not require qualification. All critical equipment in chemical production used to manufacture chemical intermediates will be qualified.

Only Pharmaceutical Development equipment that is used to manufacture drug formulations intended for use in clinical trials requires qualification. Laboratory scale manufacturing equipment in pharmaceutical development does not require qualification. All critical equipment in pharmaceutical production used for the manufacture of drug products will be qualified.

An impact assessment is the first step of qualification and is designed to determine the need for and extent of qualification of new equipment. This assessment defines the level of impact a system or component has on the product quality. Those systems with no impact or with indirect impact on product quality will be qualified and documented according to Good Engineering Practices. For equipment known to have a direct impact such as compounding tanks, autoclaves, and filling lines, a formal impact assessment is covered under IQ/OQ/PQ protocols.

Design Qualification is the first step in qualifying new equipment. It includes a review of the User Requirement Specification (URS), Piping & Instrumentation Diagrams (P&ID), functional specifications, and the GMP risk assessment. The design qualification is to be completed and approved before performing the expanded design requirements phase during which suppliers of components and parts are identified.

Installation Qualification provides documented evidence that all components listed in the overall qualification plan were installed correctly and according to design requirements such as specifications, purchase orders, contracts, and bids. The IQ phase consists of several verifications depending on the equipment and will be preceded by a Factory Acceptance Test (FAT), commissioning and Site Acceptance Test (SAT) to include; leak tests, name plate verification, and P&ID verification. Certain aspects such as manufacturer, model number, and materials of construction that are verified at the supplier during an FAT do not have to be repeated if the equipment is not dismantled or altered during shipping.

All IQ activities must be documented, verified, and approved. Documentation may include specific checklists or qualification records that refer to the specific documents that were checked. All IQ documents must be initialed and dated with a clear reference to the specific IQ checks performed. IQ must be completed, with an open items list if required, for each component before starting OQ. Signoff of all executed checks, for example checklists is acceptable evidence that the IQ was successful and that OQ can commence. Exceptions must be documented and evaluated and approved by Quality.

Operational Qualification provides documented evidence that all components are capable of operating within established limits and tolerances. Operational Qualification involves testing parameters that regulate processes or product quality to ensure they are operating within pre-defined limits. OQ verification includes a documented assessment of the proper operation of controllers, indicators, recorders, alarms, and interlocks by performing single tests on each component or integrated tests of systems such as heating/cooling testing of compounding tanks. Details of the tests to be performed are given in approved SOPs generated during equipment qualification for the specific tests to be performed. Some tests performed during IQ can be referenced in OQ without having to be repeated. For example, worst-case capacity testing of a WFI distribution system can be tested during IQ and referenced in support of OQ.

Calibration of instruments identified as critical during IQ must be verified at the beginning of Operation Qualification before commencing testing. Some calibrations performed by the supplier may be accepted without having to be repeated. For example, alarms, interlocks, and sequencing if appropriately tested and documented during FAT need not be repeated. Supplier's calibration of temperature probes if properly documented need not be repeated since they function or do-not function – with no drift.

All OQ activities must be documented, verified, and approved. Documentation may include specific checkout sheets or qualification records that refer to the specific documents that were checked. All OQ documents must be initialed and dated with a clear reference to the specific OQ checks performed. OQ must be completed, with an open items list if required, for each component before starting PQ. Signoff of all executed checks, for example checkout sheets, are acceptable evidence that the OQ was successful and that PQ can commence. Exceptions must be documented and evaluated and approved by the Quality. At the conclusion of OQ a handoff of the project occurs, transferring responsibility for the remainder of the project from engineering to the user. This handoff includes the open-items list

Performance Qualification integrates procedures, personnel, systems, and materials of a system to verify that all associated direct and indirect impact systems produce the output specified in the URS. The output may be measured as the performance of an individual component or of the integrated system. For example, the capacity of a heating/cooling system of a compounding tank may be evaluated by the temperature of a water-filled tank. Similarly, the effectiveness of an air handling system may be evaluated by evaluating the temperature, humidity, air pressure, and airborne particulates in the rooms supplied. Performance Qualification differs from Process Validation in that in PQ the system is qualified for general use. In Process Validation, the system is verified for a specific product. If the equipment is dedicated to one product, Process Validation serves as Performance Qualification of the equipment.

Equipment Cleaning Validation is related to equipment qualification and is addressed in a separate specific SOP. Cleaning validation need not be completed before the completion of the qualification of the equipment.

Computer System Validation (CSV) involves facilities, equipment, and utilities that are often controlled by automated systems. Qualification in this case must also consider the level of computerization of the system. There are two general types of automated systems:

- Stand-alone systems are computer systems which link or control multiple systems and equipment. Examples of stand-alone systems are Building Management Systems (BMS) and System Control and Data Acquisition Systems (SCADA). These systems are usually validated as independent systems with their respective Qualification Plans and Protocols.
- Integrated control systems are computer systems that control a single system or piece of equipment. Examples of integrated control systems are Programmable Logic Controllers (PLCs) that, for example control operation of an autoclave, or linear/vertical motion on filling equipment. The Validation/Qualification of these systems are usually included in the Validation/Qualification Plans and Protocols of the system or equipment.

The qualification plan or protocol defines all critical activities to be performed during qualification of a given system or equipment, including the related responsibilities for qualification and the acceptance criteria. A qualification protocol may cover all qualification steps for small projects; however, a Master Qualification Plan is usually required for larger projects. Engineering and Quality will collectively establish the level required for each qualification step according to the level of complexity of the project and provide all details in the Protocol/Plan.

Qualification Protocols/Plans will include:

- Scope of the qualification
- A detailed description of the project
- An organization chart for the project
- Qualification activities to be performed and associated responsibilities
- Change control during the project and after release of the equipment

All controls, tests, and activities to be performed during qualification must have an associated SOP or list of checks (checklists) to be performed depending on the complexity of the test. The SOP/checklists are generated by the Engineering Department. These SOP/checklists are designed to be used for multiple systems or pieces of equipment; however, unique systems may require additional SOPs. The exact SOP or checklist followed in each step of qualification will be fully documented.

Documentation of qualification activities will be in the respective checklist, otherwise a separate test plan will be established that provides for documentation of the test performed. All qualification documents must be completed according to good documentation practices, signed and dated, and must be included as part of the qualification documentation. The documentation of suppliers can be used. It has to be assured that the documentation is compliant to regulatory and internal requirements.

All qualification activities and results will be documented in a report that clearly states if the qualification was successful or not. This final report once approved serves as the formal release of the system or equipment for use in production. Depending on the complexity of the qualification there may be separate reports for or each phase of qualification.

The qualification report will include:

- An executive summary
- The scope of the qualification
- A summary of the results (tables, data)
- Appendices capturing all work in checklist and all relevant data collection
- Reference to the detailed preventive maintenance instructions for major equipment, which may be issued as working instructions.
- Comments regarding pending change requests and the open-item list
- Conclusions

Equipment requiring qualification must be released by Quality for use in GMP processes. A conditional release may be given by Quality in writing (Interim Report), pending compilation of the final report provided all testing is completed and has been reviewed. But in each case, all critical deviations must be investigated before the equipment is (conditionally) released.

Changes in general must be evaluated, documented, verified, and approved. Minor changes, however, such as replacing a mechanical component for a like-for-like type, model, etc., on a piece of equipment does not require formal verification and approval. It is documented through a work order.

During a project, the handling of changes will be defined in the qualification plan. The qualification team will identify those changes that must be approved and documented. After release of the equipment, approved change control procedures will be followed.

Qualified systems or equipment will be evaluated on periodic basis after initial qualification or requalification or last review to verify that they are still operating in a valid manner. The period must be defined internally to ensure that critical systems remain reliable before the next evaluation. The evaluation will include a review of the P&IDs, change log, and results of product performance. When no significant changes have been made, the periodic review of equipment production and maintenance logs will confirm that the system or equipment consistently produces product meeting specifications and document that requalification is not required.

23 Process Validation

Validation is an evaluation providing documented evidence that a process performs consistently and reliably, meeting predetermined processing and product specifications. The robustness of the manufacturing process, robust analytical tests, and product specifications are generally established during development. Data and/or scientific rational will be available to support the established ranges. Generally, three consecutive, successful batches are required to complete a validation study. Validation batches will be identified as such prior to the production of first validation batch. Validation protocol must be written and approved with associated procedures issued in controlled documents and approved prior to beginning the validation process. The protocol will include acceptance criteria.

The preferred approach to process validation is prospective validation. There are, however, some circumstances when either concurrent validation or retrospective validation is an acceptable practice. Prospective validation is conducted prior to distribution of either a new product or product that has been significantly changed, i.e., made under a revised manufacturing process where the revision may affect product characteristics. It has to be completed before the commercial distribution of the final drug product or drug substance.

Concurrent validation is acceptable when data from three replicate production runs cannot be generated because only a limited number of batches can be produced. Rework of an API/intermediate or limited demand of a drug product are primary examples for this approach. Prior to the completion of concurrent validation, validation batches can be released and used for commercial distribution provided they meet all validation acceptance criteria. Pre-validation batches produced prior to the concurrent validation batch(es) can be released and used for commercial distribution provided they were produced under GMP, meet all validation sampling and acceptance criteria, and there is no significant process change between the pre-validation batches and validation batches. If a concurrent validation batch fails to meet the acceptance criteria, batches that were already released for commercial distribution may be subject to recall.

Retrospective validation may be conducted for well-established processes that have been used without significant changes in raw materials, equipment, systems, facilities, or the production process. This validation approach may be used where: Critical quality attributes and critical process parameters have been identified and documented. Appropriate in-process acceptance criteria and controls have been established. No significant process/product failures attributable to causes other than operator error or equipment failures unrelated to equipment suitability have occurred. Impurity profiles have been established for the existing API. Batches selected for retrospective validation should be representative of all batches produced during the review period, including any batches that failed to meet specifications, and should be sufficient in number (10–30 consecutive batches) to demonstrate process consistency. Retained samples can be tested to obtain data to retrospectively validate a process.

Process Qualification is conducted when there is a change in process equipment or critical production materials, such as compounding ingredients that are likely to affect the process validation status. Qualification normally requires the evaluation (e.g. impurity profiles) of three batches with routine sampling to ensure that the change did not have a negative effect on product quality.

Process validation of APIs for use in clinical trials is normally not required where a single batch is produced or where process changes during development make batch replication impractical. The combination of process controls, calibration, and equipment qualification control API quality during development. Process validation should be performed also on a pilot or small-scale batche when the material is planned to be used commercially.

All critical manufacturing steps (steps including critical process parameters or steps which are defined as critical) in chemical production after introduction of the API starting material into the process will be validated. All steps from the final Intermediate on are also validated. Process validation prior to the starting material will be performed by exception according to the potential impact on the product and/ or customer (Quality Agreement).

Process validation of drug products for use in early clinical trials is normally not required, where a single batch is produced or where process changes during development make batch replication impractical. The combination of process controls, calibration, and equipment qualification controls drug product quality during development. Process validation should be performed when batches in Pharmaceutical Development are produced for commercial use, even when such batches are produced on a pilot or small scale.

All critical manufacturing steps in Pharmaceutical Production necessary to produce the drug product beginning from the first blending operation to the bulk drug product will be validated according to validated procedures. Once it has been determined that a manufacturing process requires validation, the validation coordinator gathers all information needed to prepare the validation protocol and perform the validation.

Chemical manufacturing process validations involving more than one step will begin with a Master Validation Plan (MVP) prior to the approval of the validation protocols for each individual step. This MVP describes all validation activities necessary for the validation of the manufacturing process. In a multi-step production, for example: component preparation, compounding, and filling operations, each step is considered as an independent process, i.e. each step is validated independently from the others.

Validation of individual steps in general will proceed as follows: Approve Validation Protocol with pre-determined acceptance criteria. Monitor critical process parameters. Evaluate analytical results. Approve Validation Report including all data necessary.

Equipment Qualification requires that all equipment associated with the process must be qualified and in current calibration at the time of validation. If the product is produced in a controlled environment, the HVAC equipment must be qualified or undergo qualification concurrently (for new installations) and an ongoing environmental monitoring program should be implemented.

Regulatory Dossier, Drug Master File (DMF), and other regulatory filings should include changes to the process, and analytical methods that affect the regulatory filings as a result of process changes mandating validation will be included in an update to the appropriate filing after validation but prior to release of the product to the end customer. Ensuring appropriate disposition of the Product is handled in the Change Control system.

Development Report is required for new products. It identifies the critical process parameters that are required for validation. Analytical Methods must be validated in accordance with the standard on analytical validation.

Raw Materials must be qualified including written specifications and test methods confirmation that the vendor has entered into the vendor qualification program. API Starting Materials must be tested using methods validated in accordance with the standard on analytical validation.

Process validation begins with the introduction of the API starting material into the process. All critical steps of the process as well as all steps from the Final Intermediate are included in the validation. An impurity profile comparison must be performed during process validation at those stages most likely to show impurities. Rework procedures must be validated. If the rework procedure is to be used only once. qualification including enhanced sampling and testing is required.

API milling and blending operations must be validated. That is, milled and blended lots must be tested for uniformity with respect to particle size distribution according to current specification.

Validation will cover the batch size range of the respective process. Batch size ranges are usually validated by "bracketing" using four batches at a minimum (two batches at the lower end and two batches at the upper end of the batch size range). Other approaches can be chosen but must be justified in the validation protocol. In addition to the validation samples, the normal process samples will be taken as specified by the routine sampling procedure.

All steps of bulk drug product manufacture as well as finished drug product packaging must be validated. Blending will be validated at target blending conditions. Blend uniformity of the final blend has to be tested. Validation of different strengths may be done with reduced validation effort based on worst-case considerations and on scientific rational (matrixing and bracketing).

If two or more drug substance (API) suppliers are to be used in the first validation of a finished drug product manufacturing process, a bracketing or matrixing approach could be chosen with respect to the number of validation batches (e.g. 3 batches with respect to the API of the first supplier, 1 batch with respect to the API of additional suppliers). At least 1 batch per API supplier should be required for the validation. If the supplier of the active pharmaceutical ingredient changes during the commercial life time, a new validation is necessary. The extent of the validation should be established in a risk assessment. Reduced sampling may be possible if the impact of the API on the individual process steps is not critical.

Packaging validations should be conducted for finished drug product. For each packaging configuration, at least 3 packaging validations runs should be conducted on no less than 10% of each bulk finished drug product batch. If multiple packaging configurations exist (i.e. different bottle sizes and counts, etc.) a matrix or bracketing approach can be designed to validate the packaging process.

The following criteria (if appropriate), at a minimum, must be met when conducting a process validation of chemical processes. Acceptance criteria for wet chemical processes include assay, purity, impurities, residual solvents, particle size distribution. For pharmaceutical processes acceptance criteria include relative standard deviation, content uniformity, and dissolution.

Validation of a multi-step drug product/medical device manufacturing process requires a Master Validation Plan with topics needed to be addressed at a minimum: Scope, Responsibilities, Documentation, Resources/Training, Process Description, Critical Process Parameters, Acceptance Criteria, Materials, Equipment/ Maintenance/Calibration, Cleaning, and Schedule.

A protocol must be written that details the production process. The validation report should have a summary of analytical results, comments/deviations, conclusions, and recommendations.

Process Validation (PV): Process validation is defined as the collection and evaluation of data, from the process design stage throughout production, which establishes scientific evidence that a process is capable of consistently delivering quality products. Process validation involves a series of activities taking place over the lifecycle of the product and process. FDA guidance describes the process validation activities in three stages:

Stage 1 – Process Design: Process design is the activity of defining the commercial manufacturing process that will be reflected in the master production and control records. The goal of this stage is to design a process suitable for routine commercial manufacturing that can consistently deliver a product that meets its critical quality attributes. The commercial process is defined during this stage based on knowledge gained through development and scale-up activities.

Stage 2 – Process Qualification: During this stage, the process design is confirmed as being capable of reproducible commercial manufacturing. This stage has two elements: (1) design of the facility and qualification of the equipment and utilities and (2) performance qualification (PQ). During this stage, cGMP compliant procedures must be followed and successful completion of this stage is necessary before commercial distribution. Products manufactured during this stage, if acceptable, can be released.

Stage 3 – Continued Process Verification: The goal of this third validation stage is to continually assure that the process remains in a state of control (the validated state) during commercial manufacture. A system or systems for detecting unplanned departures from the process as designed is essential to accomplish this goal. Adherence to the cGMP requirements, specifically including the collection and evaluation of information and data about the performance of the process will allow detection of process drift. The evaluation should determine whether action must be taken to prevent the process from drifting out of control (§211.180 (e)).

Process Validation Addendum: An addition to an existing signed validation document to provide additional information or data to the original validation.

Validation originator will complete a gap analysis report for validation studies in conformance to regulatory standards or requirement (i.e. software validation in conformance to 21 CFR 211, sterilization validation in accordance to ISO/ANSI/AAMI guideline, etc.)

Some validation processes are routine in nature and are well defined. To facilitate the control process, these validation processes may be written into procedures to describe in detail the validation process requirement. Validation procedure specifies the validation parameters as well as required approval signatures in a specified (approved) form. Since the validation protocol requirements are described in the procedure and summarized and approved in the form, the approved form can be attached and used for archival purposes. The final report will need to go through the approval process accordingly.

Engineering also performs Quality Risk Analysis, provides data and defines critical process parameters/steps where appropriate to ensure that process parameters used for validation are appropriate to produce a Product that meets pre-defined specifications.

Process Capability Index: the capability of a process to meet customer requirements determined by calculating the short-term process stability by determining the variations between a fixed number of samples. CpK is defined as the minimum (CPU, CPL). The CpK must meet a minimum of 1.0.

$$CPU = \frac{UCL - \mu}{3\sigma} \qquad CPL = \frac{\mu - LCL}{3\sigma}$$

where: UCL: Upper Critical Limit
LCL: Lower Critical Limit
μ: average σ: Standard Deviation

24 Change Control

Revalidation is required in the following cases: Any major change is implemented as a result of several minor changes that together are determined to potentially affect product quality. If more than 10% of the batches in a campaign fail with no assignable cause, the process must be reviewed. If appropriate, the process will be revalidated. Revalidation may also be required by the customer. In addition to the annual product review, a review of the validation status will be performed after a specified period. The following attributes should be part of the review of the validation status. The review should be documented:

- Review of current Master Batch Record/Deviations/Complaints
- Review of the last Process Validation (protocol, report)
- Review of all changes related to the manufacturing process and production equipment
- Verification, if the last process validation complies with current requirements of SOP
- Review of analytical method changes/specifications

Major Changes include:

- Equipment
- Relaxing or deleting a raw material specification that is likely to affect product quality
- Product specification change not based on process performance
- Change of the API supplier
- Process changes
- Batch size increase/decrease
- Critical parameters
- Addition/deletion of critical processing steps

Minor Changes include:

- Editorial changes to batch records
- Waste treatment/air emissions changes
- Change to "equivalent"equipment (Like-for-like)
- Changes to non-critical parameters
- Tightening of critical parameters

25 Analytical Testing of Raw Material Quality

The extent of raw material testing is determined by the manufacturer. A conservative approach would be to perform complete analysis of each lot of raw materials received. USP provides monographs for the most commonly used raw materials in the pharmaceutical industry. Analytical raw material testing governs API and excipient testing to support USP/NF, EP, BP, JP, FCC, and ACS monograph testing requirements. Testing supports formulation development of excipients and active pharmaceutical ingredients used in finished product manufacturing and supports the qualification of raw material vendors.

Raw Material Testing services should include Complete Compendia Testing (USP, EP, BP, JP), USP <467> Residual Solvent Testing, HPLC, GC and IC analyses, Spectrophotometric Analyses (FTIR, UV/VIS), Metals Analyses (AA, GFAA, ICP), Heavy Metals Testing, Ethylene Oxide and 1,4 – Dioxane Testing, Karl Fisher titration Analyses, and Wet Chemistry Analyses. Impurities and residual degradation products affect formulation quality and specification requirements to formulation stability and efficacy. Therefore, Method Development, Method Validation, Method Transfer, Extractable/Leachable Studies are of prime importance to quantify and verify raw materials with regard to Elemental Impurities.

Raw Material, Process Aid, Cleaning Agent, API, all require testing. Samples/batches of Raw Material from the new vendor are compared with 3 representative samples/batches from the existing vendor. If there are only 2 batches available, the comparison could be done using these 2 samples as an exception.

26 Product Life Cycle

Quality Risk Management (QRM) and Quality by Design (QBD) concepts are incorporated into facility and system verification efforts. At the beginning of 2011, the US FDA published an update on pharmaceutical process validation titled "Guidance for Industry – Process Validation: General Principles and Practice". Based on the principles of IGH Q8, Q9, and Q10 (Pharmaceutical Development, Quality Risk Management, and Pharmaceutical Quality Systems) the FDA guide (PVG) targeted risk-based GMP initiative. The PVG is structured on lifecycle concept: The objective of process validation is a state of ongoing control across entire product development and manufacturing lifetime. The specific architecture that the agency applies to the life cycle model is a three-stage model that begins with process design and ends only with the discontinuation of manufacture. This contains commissioning and qualification, which is referred to as stage 2a of the FDA process validation cycle. ASTM E2500-07, standard guide for specification, design, and verification of pharmaceutical and biopharmaceutical manufacturing systems and equipment is cited by FDA PVG as guidance for activities that verify that facilities, systems, and equipment are fit for their intended use as verified by Qualification.

27 Quality by Design (QbD)

The fundamental principles of PVG are that quality must be designed into a process from the beginning. It cannot be adequately ensured merely by inspection or sampling and testing

The key to defining facility and equipment quality requirements is based on the process knowledge gained during stage1 – Process Design. PVG states "This knowledge and understanding is the basis for establishing an approach to control of the manufacturing process those results in products with the desired quality attributes". FDA expects manufacturers to:

- Understand the source of variation
- Detect the presence and degree of variation
- Understand the impact of variation on the process and ultimately on product attributes
- Control the variation in a manner commensurate with the risk it represents to the process and product

DESIGN-BASED CONTROL STRATEGIES

Critical aspects of manufacturing systems are typically functions, abilities, performance or characteristics necessary for the manufacturing process and systems to ensure consistent product quality and patient safety. Failure Modes and Effects Criticality Analysis (FMECA) is an example of a type of risk assessment method that is practical in identifying specific risk and suggest controls, which is applicable to critical aspects identification. Other scientific approaches such as design of experiments might be applicable.

Assurance that manufacturing systems are fit for intended use should not rely solely upon verification after installation but be achieved by planned and structured verification approach throughout the system lifecycle.

To effectively apply QbD and QRM to the design and delivery of manufacturing facilities, equipment, and systems, a functional level of process knowledge regarding the intended use of assets must be available. Based on this knowledge, Subject Matter Experts (SMEs) can define the appropriate quality requirements, according to regulations (guides), and specific requirements affecting product quality and patient safety with respect to product knowledge, process knowledge, regulatory requirements, and company quality standards. Product and process knowledge come from earlier stages (FDA –Stage 1) of process design and development along with relevant manufacturing experience linking the engineering designs of the facilities and systems that support those requirements. Namely, the critical utilities systems including water for injection (WFI), clean steam (CSM), clean compressed gases (CCG), heating ventilation and air conditioning (HVAC), reverse osmosis water (RO), and classified clean rooms should all be governed by establishing a trail of documentation to cover: user requirement specifications (URS), machine design specifications (MDS), software design specifications (SDS), hardware design specifications (HDS), functional design specifications (FDS), design review (DR), installation, operation, and performance qualifications (IQ, OQ, PQ). Similarly, manufacturing systems should follow all the above requirements in addition to factory acceptance test (FAT), commissioning and site acceptance test (SAT), and following PQ a full equipment criticality analysis (ECA) to define gaps and equipment life cycle for sustainable compliance in producing safe and efficacious medicinal and pharmaceutical products.

ASTM E2500-07 provides the necessary high-level strategy for science and risk-based verification that facilities and systems are fit for use. In addition, commissioning and qualification methodologies should be embedded in quality systems across the design, manufacture and distribution of life sciences–regulated products. To facilitate this effort, ISPE published two guides in 2011. Science and Risk-based Approach for the Delivery of Facilities, Systems, and Equipment (FSE Guide) and Applied Risk Management for Commissioning and Qualification (ARM Guide). FSE guide emphasis is on new or flexible quality systems without significant legacy. The ARM guide focuses on embedded terminologies within established quality systems, whereby the culture is non-risk-based, and the organization quality systems maturity needs development.

FDA process validation guidance establishes a three-stage lifecycle: Stage 1–process design and development, Stage 2 is divided into 2(a) for C&Q of equipment, systems, and facilities and 2(b) to focus on risk management. The goal is to support verification approaches that systems are fit for their intended use.

GLOBAL HARMONIZATION TASK FORCE

The Global Harmonization Task force (GHTF) and US FDA process validation guidance are compared to show that they have substantial overlap in both pharmaceutical and medical device standards. Manufacturing products whose predicate regulations require process validation such as drugs, medical devices, API, biologics should incorporate FDA standards plus assure the highest quality standards as being developed in process science and validation. FDA Process Validation Guide: General Principles and Practice was finalized in January 2011. On-going Process Validation emphasizes the fact that Performance Qualification (PQ) is not the end of validation; rather it marks the start of commercial production that will require certain equipment to be requalified or revalidated periodically to ensure continuous process validation over time. Statistical analysis, statistical process control, and process capability and trending are typical markers of the requirements in process validation.

GHTF process validation standard, SG3/N99-10:2004, Quality Management Systems – Process Validation Guidance remains applicable and FDA 2011 standard explicitly stated that device firms were to follow GHTF guidelines ISO 13485:2003

– Medical devices – QMS updated to harmonize with ISO 9001:2000. This is of particular interest to a combination drug manufacturer that produces both drugs and devices, whereby there is a need to comply with both GHTF and 2011 FDA guidance standards.

The FDA process validation guidance standards expect engineering studies to be performed to determine the critical processing parameters and their operational ranges that produce an acceptable final product. The GHTF, in contrast assigns these activities to the operation qualification instead of an earlier phase, which renders it as an exploratory experiment than a rigorously defined protocol. Our practice should follow FDA standards provided that operating and alert parameters are documented to satisfy GHTF standards, and that the associated equipment is qualified to meet process specifications at established limits.

Firms that produce drugs or medical devices are required to perform studies to manage risk and employ statistical tools. Besides the reference to ISO 14971, Application of Risk Management to Medical Devices, GHTF standard references the use of fault tree analysis (FTA) and Failure Modes and Effects Analysis (FMEA) to determine which aspects of a process pose the greatest risk to product quality. FDA guidance references Design of Experiments to identify relationships between control and component inputs and process output characteristics in terms of significant interactions in the process. Proper documentation of statistical strategies in process validation is important to establish confidence in the results to ensure, "high degree of assurance" requirements (820.75) such that this predicate is not violated, which will result in an inspection to declare the entire validation effort null and void. Therefore, risk assessment and statistical requirements should be employed to ensure compliance of a validated process.

Building management systems and off-the-shelf programs that store labeling artwork and print and reconcile labels have internal software processes that function independently of the equipment being monitored and operated; as such, these type systems might warrant their own validation as part of the facilities, in contrast a PLC that was coded to operate the specific machine, for example a heat sealer is arguably an integral part of that equipment validation. Software validation of the overall system challenges ladder logic as part of equipment qualifications combined with code documentation and change control meets requirements under FDA guidance regulations.

GAMP 5 as issued by ISPE to define determinations whether software is, or is not, an integral part of equipment design. Validation plans or risk assessment documents would reference software 21 CFR part 11 (electronic records) impact. Although there are no specific requirements to separate validation efforts as a result of electronic record implications, company standards with respect to FDA part 11, ISO 13485, and EU Annex 11 tie validation of systems that process electronic records and electronic signatures (ERES) to the standard based on a separate computer system validation (CSV), which is audited separately.

Overall, the classic acceptance of three production runs to support a performance qualification has been established. FDA accepts that three production runs during process validation is a practical standard; FDA recognizes that all processes may not be defined in terms of lots or batches. Three is scant for accepting arbitrarily a successful continuous trend. On-going monitoring of process variability and trending is a regulatory expectation (usually CpK trending above 1.3). Trends in the process should be monitored to ensure the process remains within established parameters. Data collected should include relevant process trends. Data should verify that the critical quality attributes are being controlled throughout the process.

28 Sterilization

Drug Products and Materials that are used in medical and surgical applications, which are used in healthcare facilities are made of materials that are heat stable and therefore undergo heat, primarily steam, terminal sterilization. An increase in medical devices and components made of materials (e.g., plastics) that require low-temperature sterilization necessitated the use of other means of sterilization such as radiation and ethylene oxide gas, which has been used for heat- and moisture-sensitive medical devices, and ingredients. Recently, low-temperature sterilization systems (e.g., hydrogen peroxide gas plasma, peracetic acid immersion, ozone) have been developed and are being used to sterilize medical devices. This section reviews sterilization technologies for drug products and components used in healthcare.

Sterilization destroys all microorganisms on the surface of an article or in a drug-compounded fluid, or solutions to prevent disease transmission associated with the use of that item or compound. Inadequately sterilized critical components represent a high risk of transmitting pathogens. The sterility assurance level (SAL) of the product is defined as a single viable microorganism occurring on a product after sterilization. An acceptable SAL is the probability of a spore surviving was one in one million, the SAL would be 10−6. SAL is an estimate of lethality of the entire sterilization process.

Medical products that have contact with sterile tissues or fluids are considered critical. These items should be sterile when used because any microbial contamination could result in disease transmission. Besides sterile injectable drugs, other items include surgical instruments, biopsy forceps, and implanted medical devices. If these items are heat resistant, the recommended sterilization process is steam sterilization, because it has the largest margin of safety due to its reliability, consistency, and lethality.

29 Cleaning Validation

Based on FDA audits regarding establishing the basis for a scientific approach, besides coverage and recovery studies, a matrix study based on DOE (Hi-Lo; crevices; difficult to reach, internal structures; tiers; nozzle connections; etc.) to establish justifiable critical swabbing points (sampling locations) is recommended to ensure equipment repeatable cleanliness (Batch-to-batch; between-batches). This is usually covered under HACCP (Hazard Analysis Critical Control Points) Risk assessment program, which covers cleaning validation. (See Appendix X.)

30 Lyophilization

INTRODUCTION

Lyophilization is defined as a freeze-drying process that removes water from a product after it is frozen and placed under a vacuum. If the bulk drug ingredients are not stable in liquid or frozen form, lyophilization is necessary. This can be due to chemical reactions, degradation, aggregation, biological growth, or heat sensitivity. Lyophilization enables longer shelf life and product transport. Products can be stored at room temperature, which supports product stability.

The global lyophilization market for pharmaceutical and biotechnology products includes new injectable drugs, vaccines, and biological products that are formulated in a lyophilized form, and for many immunotherapy proteins and polypeptides this represents the preferred way to produce stable, biologically active products with a long shelf life.

Lyophilization involves manipulating the temperature and pressure of an aqueous or mixture solution so that the phase of the solution can move directly from the frozen state to the gaseous state without moving through the liquid phase/state (Sublimation). This is achieved by cooling the solution and lowering the pressure to below the triple point of water (the temperature and pressure at which water can exist in equilibrium in the liquid, solid, and gaseous states), which is 0.01°C. This allows for the removal of the solvent from the product without subjecting the product to intense heat, thereby preserving the active pharmaceutical ingredient (API) for both efficacy and shelf life.

There are four steps in the lyophilization cycle:

1- Freezing: Creates the ice matrix that holds the drug substance and other excipients. On the triple point, the product is in the liquid phase, the temperature is lowered, and the product is frozen.
2- Evacuation: Reduces pressure below the vapor pressure of ice. A vacuum is used to reduce the pressure in the lyophilizers to below the triple point of water.
3- Primary Drying: Adds heat to drive the sublimation of water vapor from the ice onto the freeze dryer condenser coils. The addition of heat moves the product from the frozen or solid state to the gaseous state, therefore removing "free ice."
4- Secondary Drying: Adds heat to drive off unfrozen water that is bound to the interstitial surface area of the cake.

The main advantages of lyophilization include:

- Enhanced product stability in a dry state
- Ease of aseptic processing of a liquid as opposed to a powder fill
- Removal of water without heating the product, retaining its efficacy or nutritive value
- Rapid and easy dissolution of reconstituted product

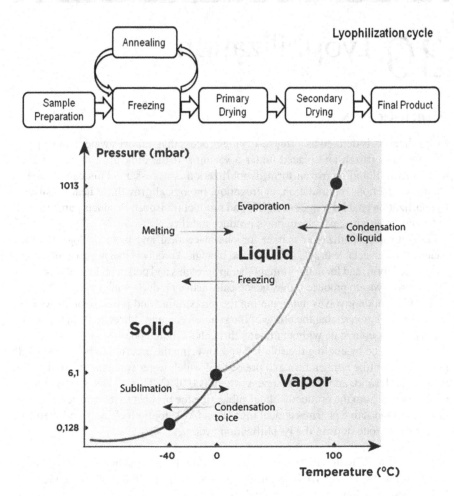

LYOPHILIZATION CYCLE DEVELOPMENT

Many parenteral pharmaceutical products, including anti-biotics and biologics, are manufactured as lyophilized products. Parameters that define the manufacturing process (compounding, filling, lyophilization, capping, inspection, packaging, and product (Lyophilized cake elegance) conditions are validated to meet product release specifications. Risk assessment is critical to identify nuanced differences between manufacturing sites or between scale-up activities that could take the process outside of the acceptable design limits, resulting in a potential failure or lost batch.

The first lyophilization run cycle parameters are based on design to evaluate how the lyophilization cycle works within installed and validated systems. The next runs are used to examine various combinations of high and low temperatures and pressures. The evaluation of the data collected from the lyophilization runs and the supporting analytical data define the ranges of the lyophilization cycle parameters. Data from the lyophilization run can include:

- Thermocouple data on shelf temperature versus product cake temperature

- Pirani gauge data, measuring vapor pressure of the sublimation process
- Pressure rise testing, used to indicate the amount of remaining un-sublimated water

Data gathered is evaluated to demonstrate that there is no change in product assay, impurity profile, residual moisture in the cake, and the reconstitution properties of the product. Additional Lyophilization runs are conducted to confirm all engineering process parameters are repeatable. One of the lyophilization runs shall incorporate a maximum pressure test to evaluate when pressure is increased so that sublimation is no longer the driving force. This test provides evidence that even if a pressure excursion occurs during the manufacturing process, there will be data to support the fact that there is no impact on the batch.

Based on the data from these first runs, a final run is designed to evaluate the scale-up of the cycle and to run the largest batch possible in the development area, prior to transferring this to the manufacturing facility. The product from this final run should represent the same operating parameters as Good Manufacturing Practice (GMP) or scale-up lyophilization product. The parameters of manufacturing for this final batch should also represent the manufacturing parameters for the GMP batch manufacturing that is acceptable for regulatory submission as an exhibit batch.

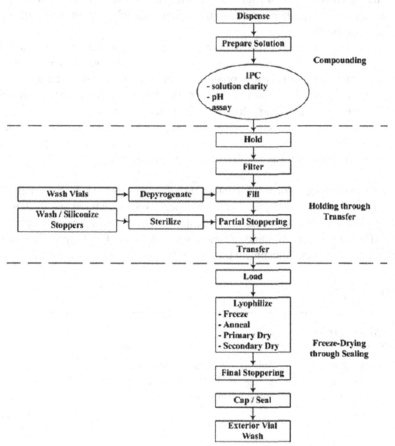

Key factors for pharma companies to consider for successfully transferring development-scale projects to GMP production include:

1. A thorough understanding of the active pharmaceutical ingredient (API), including factors such as the density, viscosity, moisture load, form (for example, amorphous or crystal), particle size, solubility, stability, and any available differential scanning calorimetry (DSC) and thermogravimetric analysis (TGA) data and material interaction compatibility studies. If the product is being supplied as a bulk drug substance, product solution handling characteristics to consider include: viscosity, solution appearance, and the possibility of solution foaming or bubbling during the filling process in manufacturing. The investigator should verify that the analytical methods have been validated.

2. Clinical program requirements for product supply chain should anticipate program supply needs, account for samples for test validation, stability programs, and clinical phase testing needs. Details of whether a placebo is needed and if this must be blinded.

3. Confirmation of whether non-GMP API is being used to start development activities and whether major changes are expected to the synthesis route. Such changes could impact the characteristics of API. For example, differences could include polymorph differences, particle size, or moisture content, or impurities.

4. If the lyophilization cycle is appropriate for the product's phase of development, life cycle optimization studies are carried out to develop well-characterized and robust approaches, including a more detailed design of experiment (DOE) studies to optimize the process within acceptable specification limits including allowances for process variations.

5. The need for product elegance, which may not be an issue for Phase 1 studies in one region but may be a major factor if the product is to be used in another region affected by differential pressures and temperatures.

Based on experience with lyophilization development projects, it is clear that success in lyophilization transfer and scale-up depends on a structured approach for both small- and large-molecule biopharmaceutical products, which include extensive details about APIs or bulk drug substance characteristics. A well-defined approach toward the development, scale-up, or technical transfer of the process will mitigate risk.

REFERENCES

Standards & Reference Documents	Title
ISPE Baseline Guide	Volume 4, Water and Steam Systems
ISPE Baseline Guide	Commissioning and Qualification
ASME BPE 2019	Bioprocessing Equipment

(Continued)

Standards & Reference Documents	Title
USP <1231>	Water for Pharmaceutical Purposes
Title 21 Code of Federal Regulations (21 CFR)	Parts 210, 211 & 820
ICH Q9	Quality Risk Management
EU GMP Annex 15, PIC/S Annex 15	Qualification and Validation
ASTM E2500	Standard Guide for Specification, Design, and Verification of Pharmaceutical and Biopharmaceutical Manufacturing Systems and Equipment

References

FOR MANUFACTURING OF API:

- ICH Q7 Good Manufacturing Practice Guidance for Active Pharmaceutical Ingredients (APIs), 1 April 2003
- FDA's Guide to inspection of Bulk Pharmaceutical Chemicals, 1985

FOR MANUFACTURING OF DRUG PRODUCT:

- The Code of Federal regulations, 21 CFR Parts 210 & 211
- Pharmaceutical Inspection Convention, PIC-GMP
- EU GMP Guidelines

US FDA, Guidance for Industry – Process Validation: General Principles and Practices, www.fda.gov

ICH Q8 (R2) Pharmaceutical Development, ICH Q9 Quality Risk Management, ICH Q10 Pharmaceutical Quality Systems, www.ich.org

ASTM E2500-07, Standard Guide for Specifications, Design, and Verification Pharmaceutical and Biopharmaceutical Manufacturing Systems and Equipment, ASTM International, www.astm.org

US FDA, Pharmaceutical cGMP for the 21st Century – A Risk –Based Approach, www.fda.gov

21 CFR Part 211, Subpart C, Buildings and Facilities, www.fda.gov

ISPE Guide: Science and Risk-Based Approach for Delivery of Facilities, Systems, and Equipment, International Society of pharmaceutical Engineering (ISPE), June 2011, www.ispe.org

ISPE Good Practice Guide: Applied Risk Management for Commissioning and Qualification, International Society of pharmaceutical Engineering (ISPE), First Edition, October 2011, www.ispe.org

FDA 21CFR Parts 210 and 211

FDA Guide to Inspections Validation of Cleaning Processes, July 1993

Eudralex Volume 4 Medicinal Products for Human and Veterinary Use

EMA Guideline on setting health based exposure limits for use in risk identification in the manufacture of different medicinal products in shared facilities, November 2014

PDA Technical Report No. 29 (Revised 2012), "Points to Consider for Cleaning Validation"

PDA Technical Report No. 49, "Points to Consider for Biotechnology Cleaning Validation"

ICH Harmonized Tripartite Guideline; Good Manufacturing Practice Guide for Active Pharmaceutical Ingredients, Q7, November 2000

ICH Harmonized Tripartite Guideline; Validation of Analytical Procedures: Text and Methodology, Q2(R1), November 2005

ISPE Risk-Based Manufacture of Pharmaceutical Products – Baseline Pharmaceutical Engineering guide, September 2010

APIC GUIDANCE ON ASPECTS OF CLEANING VALIDATION IN ACTIVE PHARMACEUTICAL INGREDIENT PLANTS, May 2014.

Glossary

Design Qualification (DQ): The documented verification that the proposed design of the facilities, systems, and equipment is suitable for the intended purpose.

Installation qualification (IQ): The documented verification that the facilities, systems, and equipment as installed or modified, comply with the approved design and manufactures' recommendations.

Operations qualification (OQ): The documented verification that the facilities, systems, and equipment as installed or modified, perform as intended throughout the anticipated operation ranges.

Performance qualification (PQ): The documented verification that the facilities, systems, and equipment as connected together, can perform effectively and reproducibly, based on the approved process method and product specification.

Process validation: The documented evidence that the process operated within established parameters can perform effectively and reproducibly to produce a medicinal product meeting its predetermined specifications and quality attributes.

FDA

Guideline on General Principles of Process Validation: Validation—Establishing documented evidence which provides a high degree of assurance that a specific process will consistently produce a product meeting its predetermined specifications and quality attributes.

Guidance for Industry: Quality Systems Approach to Pharmaceutical Current Good Manufacturing Practice Regulations: With proper design and reliable mechanisms to transfer process knowledge from development to commercial production, a manufacturer should be able to validate the manufacturing process. In a quality system, process validation provides initial proof, through commercial batch manufacture, that the design of the process produces the intended product quality.

EU GMP GUIDE ANNEX 15

Chapter Process validation § 20: The requirements and principles outlined in this chapter are applicable to the manufacture of pharmaceutical dosage forms. They cover initial validation of new processes, subsequent validation of modified processes, and revalidation.

Site Acceptance Testing (SAT): It is a documented testing similar in nature to that of the Factory Acceptance Test; however, testing is conducted in the facility (site) on a much scaled-down version. SAT is conducted to ensure that no damage had occurred to the equipment in transit and that the unit functions properly. Determination of units for product OQ is specified by product specifications only.

Nomenclature

Abbreviations	Description
FDA	Food & Drug Admin
ISPE	International Society of Pharmaceutical Engineers
EMA	European Medicines Agency
MHRA	Medicines & Healthcare Products Regulatory Agency (UK)
ICH	International Council for Harmonization
GMP	Good Manufacturing Practice (GLP – Laboratory; GEP – Engineering; GXP – generic)
PAI	Pre-Approval Inspection (by FDA)
ISO	International Organization for Standardization
EU	European Union
WFI	Water for Injection (pharmaceutical high grade, highly purified water) (Intravenous injection water)
PFS	Prefilled Syringe
IV	Intravenous Injection (Infusion)
MES	Manufacturing Execution Systems
SOP	Standard Operating Procedure
WHO	World Health Organization
PIC/S	Pharmaceutical Inspection Cooperation Scheme
COFEPRIS	Federal Commission for Protection against Sanitary Risk - Mexico
ANVISA	National Health Surveillance Agency – Brazil
Ø	Diameter
CFU	Colony Forming Unit (microbes)
EN ISO	International standard adopted by European Union
CIP	Clean in Place system for cleaning and sanitization
SIP	Steam in Place system for sterilization
API	Active Pharmaceutical Ingredient
LAL	Limulus Amebocyte Lysate testing (Bacterial endotoxin testing)
PLC	Programmable Logic Controller
HEPA	High Efficiency Particulate Air – Filter
BFS	Blow Fill Seal
FMEA	Failure Mode Effects Analysis
FMECA	Failure Modes Effects Criticality Analysis
HACCP	Hazard Analysis Critical Control Points
DOE	Design of Experiments
QA/QC	Quality Assurance/Quality Control
CAPA	Corrective Action Preventive Action
CEDR	Center for Evaluation of Drugs & Research – FDA
CEBR	Center for Biologics Evaluation & Research – FDA
UV	Ultraviolet light
CAR-T	CAR-T cell therapy – T immune system cells taken from patient blood and changed in lab. The special receptor chimeric antigen receptor (CAR) gene, which binds to a protein on the patient cancer cell, is added in the lab. CAR-T cells are grown in the lab and then administered to patients by infusion.

Bibliography

1. Cleaning Validation: Practical Compliance Solutions for Pharmaceutical Manufacturing, Volume 1 By Destin LeBlanc
2. Cleaning Validation: Practical Compliance Solutions for Pharmaceutical Manufacturing, Volume 2, By Destin LeBlanc
3. Cleaning Validation: Practical Compliance Solutions for Pharmaceutical Manufacturing, Volume 3, By Destin LeBlanc
4. Cleaning Validation: Practical Compliance Solutions for Pharmaceutical Manufacturing, Volume 4, By Destin LeBlanc
5. Cleaning and Cleaning Validation, Volumes 1 and 2, Edited by Paul Pluta
6. Cleaning and Cleaning Validation: A Biotechnology Perspective
7. Cleanroom Microbiology, By Tim Sandle, R. VijayaKumar
8. Cold Chain Chronicles: A practitioners outside-the-box perspectives on the importance of temperature-sensitive drug stewardship, By Kevin O`Donnell
9. Combination Products: Implementation of cGMP Requirements, Edited by Lisa Hornback
10. Computerized Systems in the Modern Laboratory: A Practical Guide, By Joseph Liscouski
11. Confronting Variability: A Framework for Risk Assessment, Edited by Diane Petitti, Richard Prince
12. Container/Closure Integrity Assessment A Compilation of Papers from the PDA Journal of Pharmaceutical Science and Technology
13. Contamination Control in Healthcare Product Manufacturing, Volume 1 Edited by Russell Madsen, Jeanne Moldenhauer
14. Contamination Control in Healthcare Product Manufacturing, Volume 2, Edited by Russell Madsen, Jeanne Moldenhauer
15. Contamination Control in Healthcare Product Manufacturing, Volume 3, Edited by Russell Madsen, Jeanne Moldenhauer
16. Contamination Control in Healthcare Product Manufacturing, Volume 4, Edited by Russell Madsen, Jeanne Moldenhauer
17. Contamination Control in Healthcare Product Manufacturing, Volume 5, Edited by Russell Madsen, Jeanne Moldenhauer
18. Contamination Prevention for Nonsterile Pharmaceutical Manufacturing, By Andrew Dick
19. Effective Implementation of Audit Programs, By Miguel Montalvo
20. Encyclopedia of Rapid Microbiol Methods, Volume 4, Edited by Michael Miller
21. Encyclopedia of Rapid Microbiological Methods, Volumes 1, 2 and 3, Edited by Michael Miller
22. Environmental Monitoring, Edited by Jeanne Moldenhauer

23. Environmental Monitoring: A Comprehensive Handbook, Volume 1, 2 and 3, Edited by Jeanne Moldenhauer
24. Environmental Monitoring: A Comprehensive Handbook, Volume 4, Edited by Jeanne Moldenhauer
25. Environmental Monitoring: A Comprehensive Handbook, Volume 5, Edited by Jeanne Moldenhauer
26. Environmental Monitoring: A Comprehensive Handbook, Volume 6, Edited by Jeanne Moldenhauer
27. Environmental Monitoring: A Comprehensive Handbook, Volume 7, Edited by Jeanne Moldenhauer
28. Environmental Monitoring: A Comprehensive Handbook, Volume 8, Edited by Jeanne Moldenhauer
29. Environmental Monitoring: A Comprehensive Handbook, Volumes 4, 5, 6 and 7 Edited by Jeanne Moldenhauer
30. Ethylene Oxide Sterilization Validation and Routine Operations Handbook By Anne Booth
31. FDA Warning Letters: Analysis and Guidance, By Jeanne Moldenhauer
32. Fungi: A Handbook for Life Science Manufacturers and Researchers, Edited by Jeanne Moldenhauer
33. GMP in Practice: Regulatory Expectations for the Pharmaceutical Industry, Fifth Edition, By Tim Sandle, James Vesper
34. Global Sterile Manufacturing Regulatory Guidance Comparison
35. Good Distribution Practice: A Handbook for Healthcare Manufacturers and Suppliers, Volume 1, Edited by Siegfried Schmitt
36. Good Distribution Practice: A Handbook for Healthcare Manufacturers and Suppliers, Volume 2, Edited by Siegfried Schmitt
37. Hosting a Compliance Inspection By Janet Gough
38. Introduction to Environmental Monitoring in Pharmaceutical Areas By Michael Jahnke
39. Laboratory Design: Establishing the Facility and Management Structure Edited by Scott Sutton
40. Lessons of Failure: When Things Go Wrong In Pharmaceutical Manufacturing Edited by Maik Jornitz, Russell Madsen
41. Lifecycle Risk Management for Healthcare Products: From Research Through Disposal, Edited by Edwin Bills, Stan Mastrangelo
42. Media Fill Validation Environmental Monitoring During Aseptic Processing By Michael Jahnke
43. Method Development and Validation for the Pharmaceutical Microbiologist, By Crystal Booth
44. Microbial Control and Identification: Strategies Methods Applications, Edited by Mary Griffin, Dona Reber
45. Microbial Identification: The Keys to a Successful Program, Edited by Mary Griffin, Dona Reber
46. Microbial Risk Assessment in Pharmaceutical Clean Rooms (single user digital version), By Bengt Ljungqvist, Berit Reinmueller

47. Microbial Risk and Investigations, Edited by Karen McCullough, Jeanne Moldenhauer
48. Microbiological Culture Media: A Complete Guide for Pharmaceutical and Healthcare Manufacturers, By Tim Sandle
49. Microbiological Monitoring of Pharmaceutical Process Water By Michael Jahnke
50. Microbiology in Pharmaceutical Manufacturing, Second Edition, Volumes 1 and 2, Edited by Richard Prince
51. Microbiology, and Engineering of Sterilization Processes, Twelfth Edition 2007, By Irving Pflug
52. PDA Technical Series: Endotoxin Analysis and Risk Management
53. PDA Technical Series: Pharmaceutical Glass PDF Single user
54. Pharmaceutical Contamination Control: Practical Strategies for Compliance, Edited by Nigel Halls
55. Pharmaceutical Legislation of the European Union, Japan and the United States of America - An Overview, By Denyse Baker, Joanne Hawana, Takayoshi Matsumura, Edited by Barbara Jentges
56. Pharmaceutical Outsourcing: Quality Management and Project Delivery Edited by Trevor Deeks, Karen Ginsbury, Susan Schniepp
57. Pharmaceutical Quality Edited by Richard Prince
58. Pharmaceutical Quality Control Microbiology: A Guidebook to the Basics By Scott Sutton
59. Phase Appropriate GMP for Biological Processes: Pre-clinical to Commercial Production Edited by Trevor Deeks
60. Practical Aseptic Processing Fill and Finish, Volumes 1 and 2 Edited by Jack Lysfjord
61. Quality By Design: Putting Theory Into Practice Edited by Siegfried Schmitt
62. Radiation Sterilization: Validation and Routine Operations Handbook By Anne Booth
63. Rapid Sterility Testing Edited by Jeanne Moldenhauer
64. Recalls of Pharmaceutical Products: Eliminating Contamination and Adulteration Causes By Tim Sandle
65. Recent Warning Letters Review for Preparation of a Non-Sterile Processing Inspection, Volume 2 By Jeanne Moldenhauer
66. Recent Warning Letters Review for Preparation of an Aseptic Processing Inspection, Volume 1 By Jeanne Moldenhauer
67. Risk Assessment and Management for Healthcare Manufacturing: Practical Tips and Case Studies By Tim Sandle
68. Risk Assessment and Risk Management in the Pharmaceutical Industry: Clear and Simple By James Vesper
69. Risk-Based Compliance Handbook By Siegfried Schmitt
70. Risk-Based Software Validation By Janet Gough, David Nettleton
71. Root Cause Investigations for CAPA: Clear and Simple By James Vesper

72. SOPs Clear and Simple: For Healthcare Manufacturers By Brian Matye, Jeanne Moldenhauer, Susan Schniepp
73. Software as a Service (SaaS): Risk-Based Validation with Time-Saving Templates By Janet Gough, David Nettleton
74. Square Root of (N) Sampling Plans: Procedures and Tables for Inspection of Quality Attributes By Joyce Torbeck, Lynn Torbeck
75. Steam Sterilization: A Practitioner's Guide Edited by Jeanne Moldenhauer
76. Sterility Testing of Pharmaceutical Products By Tim Sandle
77. Systems Based Inspection for Pharmaceutical Manufacturers Edited by Jeanne Moldenhauer
78. Technology and Knowledge Transfer: Keys to Successful Implementation and Management Edited by Mark Gibson, Siegfried Schmitt
79. The Bacterial Endotoxins Test: A Practical Guide By Karen McCullough
80. The External Quality Audit By Janet Gough, Monica Grimaldi
81. The Internal Quality Audit By Janet Gough, Monica Grimaldi
82. Thermal Validation in Moist Heat Sterilization Edited by Jeanne Moldenhauer
83. Torbeck's Statistical Cookbook for Scientists and Engineers By Lynn Torbeck
84. Trend and Out-of-Trend Analysis for Pharmaceutical Quality and Manufacturing Using Minitab By Lynn Torbeck
85. Validating Enterprise Systems: A Practical Guide By David Stokes
86. Validation Master Plan: The Streetwise Downtown Guide By Trevor Deeks
87. Validation by Design: The Statistical Handbook for Pharmaceutical Process Validation By Lynn Torbeck
88. Validation of Analytical Methods for Biopharmaceuticals: A Guide to Risk-Based Validation and Implementation Strategies By Stephan Krause
89. Visual Inspection and Particulate Control By Scott Aldrich, Roy Cherris, John Shabushnig
90. Why Life Science Manufacturers Do What They Do in Development, Formulation, Production and Quality: A History By Lynn Torbeck

Appendix I: System-Level Impact Assessment

Criteria for System Classification are presented in the following table with seven questions. A YES response to at least one question indicates that the system is a Direct Impact system; otherwise it is classified as No Impact System.

Question Number	Question	Answer (Y/N)
Q1	Does the system contain critical attributes or critical design elements or perform functions that serve to meet one or more process requirements (CQAs) including CPPs?	
Q2	Does the system have direct contact with the product or production stream and does such contact have the potential to impact the final product quality or pose a risk to the patient?	
Q3	Does the system provide an excipient or produce an ingredient or solvent and could the quality of this substance impact the final product quality or pose a risk to the patient?	
Q4	Is the system used in cleaning, sanitizing, or sterilizing, and could malfunction of the system result in failure to adequately clean, sanitize, or sterilize such that a risk to the patient would result?	
Q5	Does the system establish a proper environment for the manufacturing process and could failure of the system to function properly pose a risk to the patient?	
Q6	Does the system use, produce, process, or store data used to accept or reject product, CPP's or electronic records subject to 21 CFR Part 11 and EU GMP Vol.4, Annex 11 or the local equivalent?	
Q7	Does the system provide product identification information without independent verification or is the system used to verify this information?	

Appendix II: Instrument Criticality Assessment

Criteria for the evaluation of criticality for instrument are presented in the following table with seven questions. Each instrument should be evaluated for criticality and calibration frequency adjusted for each classification.

Question Number	Question	Answer (Y/N)
Q1	Is the instrument used for the adjustment or calibration of critical components, such as a secondary/working test standard?	
Q2	Is the instrument or device used to monitor or control a critical process parameter?	
Q3	Will the failure of the instrument have a direct effect on product quality, or will the failure be the cause of a process to go into a hold or abort status which would impact product quality?	
Q4	Is information from this instrument recoded as part of the product batch record or lot release?	
Q5	Is the instrument used to create or preserve quality?	
Q6	Does the instrument measure a critical safety parameter?	
If any of the above questions are answered **YES**, then the instrument is Critical; answer Question 7 (n/a).		
If all the above questions are answered **NO**, then answer Question 7		
Q7	Is this a secondary device that does not have an impact on the process and/or product, is not used for data collection but it is a back-up safety device to a critical instrument?	
If the question above is answered **YES**, then the instrument is Non-Critical		

Appendix III: Required Testing (ISO 14644-2)

ISO 146144-1 (Federal Standard 209E)	Average Airflow Velocity m/s (ft/min)	Air Changes Per Hour	Ceiling Coverage
ISO 8 (Class 100,000)	0.005–0.041 (1–8)	5–48	5–15%
ISO 7 (Class 10,000)	0.051–0.076 (10–5)	60–90	15–20%
ISO 6 (Class 1,000)	0.127–0.203 (25–40)	150–240	25–40%
ISO 5 (Class 100)	0.203–0.406 (40–80)	240–480	35–70%
ISO 4 (Class 10)	0.254–0.457 (50–90)	300–540	50–90%
ISO 3 (Class 1)	0.305–0.457 (60–90)	360–540	60–100%
ISO 1–2	0.305–0.508 (60–100)	360–600	80–100%

FS Cleanroom Class	ISO Equivalent Class	Air Change Rate, ACH
1	ISO 3	360–540
10	ISO 4	300–540
100	ISO 5	240–480
1,000	ISO 6	150–240
10,000	ISO 7	60–90
100,000	ISO 8	5–48

Schedule of Tests to Demonstrate Continuing Compliance

Test Parameter	Class	Maximum Time Interval	Test Procedure
Particle Count Test	<= ISO 5	6 Months	ISO 14644-1 Annex A
	> ISO 5	12 Months	
Air Pressure Difference	All Classes	12 Months	ISO 14644-1 Annex B5
Airflow	All Classes	12 Months	ISO 14644-1 Annex B4

Appendix IV: Contract Manufacturing Process Steps

TECHNOLOGY TRANSFER CLINICAL DEVELOPMENT

- API Risk Assessment and Management

Collect all necessary data and information at the proposal stage to properly identify all step-by-step processes necessary to define safe handling and exposure levels, characterization, manufacturing availability, sterile conditions (filtration, gamma, ETO, etc.), validation of test methods, cost, storage conditions:

- Test Methods Transfer

Identify all test methods associated with customer clinical development projects. Quantify all tests that require additional development/validation. Quantify all straightforward methods for transfer. Identify types of methods (USP, HPLS, Titration, GC, analytical chemistry, etc.).

- Clinical Product Development – Process Transfer

Streamline new materials numbers approval. Plan all materials receipts (API, excipients, filters, components, etc.). Ensure that each project manager develops a project plan/schedule and communicates milestones to project team. Timely written protocols and execution are key to successful completion of reports for revenue generation. Plan lead times should cover all potential pitfalls to avoid deviations. Update forecast to reflect practical feasibility of task achievements and risk level.

- Manufacturing Engineering – Process Design and Development

Review and evaluate all manufacturing systems to cover component preparations, compounding and fill/finish operations including all manufacturing equipment. Review and evaluate all facilities clean rooms, and transfer rooms. Review and evaluate all critical utility systems for upgrades to ensure reliability. Develop capital appropriation plans to acquire necessary equipment, instruments, and machinery to ensure success of the manufacturing apparatus. Manage Capex for successful cost optimization and successful implementation of projects in lab systems, IT, manufacturing, and facilities

Appendix V: Equipment Qualification

Qualification data sheet: A controlled document that is part of the protocol used to record the results of tests performed during qualification and added to the final qualification report.

Design Qualification (DQ): Documented verification that the proposed design of the facilities, equipment, or systems is suitable for the intended purpose.

Direct Impact System: A system that is expected to have a direct impact on product quality. These systems are designed and qualified according to Good Engineering Practices and current Equipment Qualification requirements.

Equipment Qualification: Documented evidence that a piece of equipment or system can perform effectively and consistently against predetermined requirements. Equipment qualification generally includes design qualification (DQ), installation qualification (IQ), operation qualification (OQ), and performance qualification (PQ).

Factory Acceptance Test: Inspection and static and/or dynamic testing of systems or major system components to support the qualification of an equipment/system conducted and documented at the supplier's site.

Good Engineering Practices (GEP): Established engineering methods and standards that are applied throughout the project lifecycle to deliver appropriate, cost-effective solutions.

Hazard and Operability Study (HAZOP): The process of systematically reviewing a facility, system, or process to identify potential safety concerns.

Impact Assessment: The process of evaluating the impact of the operating, controlling, alarming, and failure conditions of a system on the quality of a product.

Indirect Impact System: A system that is not expected to have a direct impact on product quality. Indirect impact systems typically support Direct Impact Systems and are designed and commissioned according to Good Engineering Practices only.

Installation Qualifica,tion (IQ): Documented verification that the equipment or system as installed or modified complies with the approved design, the manufacturer's recommendations, and the user's requirements.

Master Qualification Plan (MQP): A high-level document which establishes an umbrella qualification plan for the entire project and is used to guide the project team in resource and technical planning.

No Impact System: A system that has no direct or indirect impact on product quality. These systems are designed and qualified according to Good Engineering Practices only.

Operational Qualification (OQ): Documented verification that the equipment or system as installed or modified operates as intended throughout the anticipated operating ranges.

Performance Qualification (PQ): Documented verification that the equipment and associated ancillary systems as a whole, perform effectively and reproducibly based on approved processes and specifications utilizing actual production materials or manufacturing operating conditions.

GMP Risk Assessment: A documented assessment that identifies specific aspects that are to be qualified based on the design and specifications of the equipment or facility.

User Requirements Specification (URS): A description of the requirements for the equipment or facility in terms of product(s) to be manufactured and conditions to be met.

ISPE Pharmaceutical Engineering Baseline Guide, Commissioning and Qualification, Volume 5

ISPE Pharmaceutical Engineering Good Practice Guide (GPC), Applied Risk management for Commissioning and Qualification

The US Code of Federal Regulations, 21 CFR Parts 210 & 211

The Pharmaceutical Inspection Convention, PIC-GMP

The EU GMP Guideline, Annex 1, 9, 11, 15, 15b

Appendix VI: API terms

Active Pharmaceutical Ingredient (API): Any substance or mixture of substances, intended for use in the manufacture of a drug product and that, when used in the production of a drug, becomes an active ingredient of the drug product.

API Starting Material (API SM): A raw material, intermediate, or an API that is used in the production of an API and that is incorporated as a significant fragment into the structure of the API. GMPs are applied from the introduction of the API starting material into the process.

Critical Process Parameter: A parameter whose value has a direct and measurable impact on the quality of the product.

Critical Quality Attribute: All attributes of the API that ensure that the drug product in which the API is used is safe and effective. These are ensured by the API specifications and GMP controls.

Drug product: The dosage form in the final immediate packaging intended for marketing. Excipients: Substances which are introduced in the manufacturing process and become part of the drug product.

Intermediate (I): A material produced during the processing steps to an API, which must undergo further molecular change before it becomes an API. The Final Intermediate (FI) is one (1) molecular change away from the API.

"Key" Parameter: A parameter in pharmaceutical manufacturing, which has an influence on the properties of the material within the appropriate process step. "Key" parameters will be treated as critical parameters as specified.

Master Validation Plan (MVP): A high-level document which establishes an umbrella validation plan for the entire process and is used to guide the project team in resource and planning.

Mother Liquor: The residual liquid that remains after the crystallization or isolation processes. It may be used for further processing.

Mother Liquor Recovery Cycle: The mother liquor recovery cycle is the manufacture of a series of batches where the first batch uses virgin materials and the subsequent batches have the mother liquor from the previous batch added to them in order to recover the product from the mother liquor. The series of batches that start with the virgin material and end with the last recycle of the mother liquor is defined as a Mother Liquor Recovery Cycle.

Appendix VII: Impurities – FDA Directive

DEPARTMENT OF HEALTH AND HUMAN SERVICES

**Food and Drug Administration
Silver Spring, MD 20993**

GENERAL ADVICE

Dear Sir or Madam:

This letter is to inform applicants with an approved or pending application for an angiotensin II receptor blockers (ARB) drug product (DP), as well as holders of related drug master files (DMFs) of FDA concerns related to the presence of one or more toxic impurities in some ARB drugs. This general advice letter summarizes FDA findings to date and provides recommended actions to take to ensure that your drug product, drug substance/active pharmaceutical ingredient (API), and raw materials are absent of these impurities or below our recommended limit.

BACKGROUND

In June 2018, FDA was informed of the presence of an impurity, identified as N-Nitrosodimethylamine (NDMA), from one valsartan API producer. Since then, FDA has determined that other types of nitrosamine compounds, e.g., N-Nitrosodiethylamine (NDEA), are present at unacceptable levels in APIs from multiple API producers of valsartan and other drugs in the ARB class. FDA has and will continue to provide periodic updates on this problem on its website (https://www.fda.gov/drugs/drugsafety/ucm613916.htm). FDA continues to collaborate with other drug regulatory agencies around the world to address this problem, and FDA is committed to working with manufacturers and applicants impacted by this problem to bring this to a full and timely resolution. FDA and other regulators, as well as many manufacturers, have developed and are using methods validated to detect and quantify a variety of nitrosamine impurities. FDA has posted these methods on its website. FDA has also issued Information Request letters to application and DMF holders to request specific information and samples of drugs thought to be at risk for the presence of a nitrosamine impurity.

Nitrosamine compounds are potent genotoxic carcinogens in several nonclinical species and are classified as probable human carcinogens by the International Agency for Research on Cancer (IARC). In fact, "N-nitroso" compounds are identified as a "cohort of concern" in internationally harmonized guidance, ICH M7, *Assessment and Control of DNA Reactive (Mutagenic) Impurities in Pharmaceuticals To Limit Potential Carcinogenic Risk.* ICH M7 recommends that known mutagenic carcinogens, such as nitrosamines, be controlled at or below the acceptable cancer risk level. Due to their known potent carcinogenic effects, and because it is feasible to limit these impurities by taking reasonable steps to prevent or eliminate their presence, FDA has determined that there is no acceptable specification for nitrosamines in ARB API and DP. Therefore, FDA advises that nitrosamines should be absent (i.e., not detectable as described below) from ARB API and ARB drug products. As an initial measure, FDA published "interim acceptable limits" for these nitrosamine impurities in ARBs. ARB DS or DP with levels of impurities exceeding these interim limits were recommended for recall from the market. FDA has used the interim limits to guide immediate decision-making for product recalls to balance the risks of potential long-term carcinogenic risk and disruption to clinical management of patients' hypertension and heart failure. FDA is now seeking the information outlined below to ensure that ARB API and DP entering the marketplace have no detectable nitrosamines.

Recent information gathered by FDA suggests several general causes of the presence of a nitrosamine impurity in ARB APIs. First, we now know that nitrosamine impurities can form during API processing under certain processing conditions and in the presence of some types of raw materials and starting materials. These materials include intermediates that are not purged in subsequent steps of the API process. A second cause appears to be from the use of contaminated raw materials used in the manufacturing process. Recovered materials, such as recovered solvents and catalysts, may pose a risk for nitrosamine formation due to the presence of amines in the solvents or catalysts sent for recovery and the subsequent quenching of these materials with nitrous acid to destroy residual azide without adequate removal. Independent recovery facilities may co-mingle solvents/catalysts from various customers or not perform adequate cleaning of equipment between customers. A similar cause may be from contaminated starting materials, including intermediates supplied by a vendor, that use processing methods or raw materials causing formation of nitrosamines in their material. Contamination from vendor-sourced raw materials and starting materials/intermediates is particularly challenging because an API producer whose process is not capable of forming a nitrosamine compound may not realize their process is at risk to the presence of such impurities. FDA is aware that some ARB producers have identified a nitrosamine in their finished API, even though they are using processes incapable of forming a nitrosamine impurity.

The multiple causes listed above can occur in the same API process. The typical tests for API purity, identity, and known impurities are unlikely to detect the presence of a nitrosamine impurity. Further, each failure mode could result in different nitrosamines, different amounts, or undetectable amounts of nitrosamine impurities in different batches from the same process and API producer.

ACCORDINGLY, **FDA** ADVISES:

1. ARB DP manufacturers test representative samples of each drug product batch, or alternatively a representative sample of each API lot used in each drug product batch; they have produced for the US market to determine whether any contain a detectable amount (defined below) of a nitrosamine impurity. Testing should include DP batches already distributed that have not yet reached their labeled expiration date as well as those not yet distributed by the DP manufacturer. Any DP batch already in distribution, as of the date of this letter, with a nitrosamine level that exceeds the FDA published interim acceptable limit should be recalled, if distributed, or quarantined pending appropriate disposition if not distributed. Any DP product batch found to contain a detectable nitrosamine impurity that is below the interim acceptable limit should not be released by the DP manufacturer for distribution unless FDA agrees that distribution is warranted to prevent or mitigate a shortage of a medically necessary drug (or if for export from the US, a shortage determination was made by the importing country's national regulatory agency). ARB DP manufacturers test representative samples of each API batch in their possession to demonstrate the absence of nitrosamines prior to use in DP manufacturing. In addition, DP manufacturers should test each API lot received from each supplier before releasing the API for use until the DP manufacturer has verified that the supplier can consistently produce API without a detectable nitrosamine (as defined below) in accordance with the CGMP regulations at 21 CFR 211 subpart E; see also FDA guidance for industry, ICH Q10, *Pharmaceutical Quality System.*

2. ARB applicants (or DP manufacturers on their behalf) report to FDA a finding of detectable nitrosamine in either an API lot or in a DP batch whether or not distributed. The Field Alert Report (FAR) regulation for ANDAs and NDAs (21 CFR 314.81) requires such reports for distributed batches. If not distributed, reporting a finding of detectable nitrosamine to the FAR system will assist FDA in timely remediation. For instructions on submitting an FAR, please refer to the draft guidance document *Field Alert Report Submission: Questions and Answers Guidance for Industry* at https://www.fda.gov/downloads/Drugs/GuidanceComplianceRegulatoryInformation/Guidances/UCM613753.pdf.

3. Applicants report to FDA a summary of the testing performed, as requested above, for the presence of any nitrosamine impurities in batches distributed in the US or exported from the US that are within their labeled expiration, even if recalled. We request a table be submitted to each application, if not already provided in previous correspondence to the application, with the following information for each batch number that is sampled and tested: product name (identify whether API or DP batch)
   * labeled strength (if DP batch)
   * date of manufacture

- labeled expiration date
- name of test method
- amount and type of nitrosamine detected, if any, or "none detected."

4. Data should be submitted to the application as a "General Correspondence;" the words "nitrosamine-related" should be prominently displayed on the cover letter.

5. Applicants of pending ARB applications should provide a written statement as General Correspondence declaring that the API supplier provides API that does not contain any detectable nitrosamine impurities. No further action in response to this letter is needed if this information has already been submitted to the application or referenced DMF.

6. ARB API producers test representative samples of each batch of an ARB API to determine whether any contain detectable nitrosamine impurities. Testing should include API batches distributed and within expiry, labeled with a 'retest by' date, and those not yet distributed. Any API batch containing a nitrosamine impurity above the interim acceptable limits should be recalled, if distributed, or dispositioned as not suitable for use in DP intended for the US market. If detected below the interim acceptable limit, the batch should not be distributed for use in DP intended for the US market unless FDA agrees that such use is warranted to prevent or mitigate a US shortage of a medically necessary drug.

7. API batches may be reprocessed, reworked, and/or reconditioned to be rendered absent of a detectable nitrosamine impurity as provided for in existing policies for amending or supplementing and controlling such operations. If a batch is found to contain nitrosamines and is reprocessed or reworked in any way, this should be reported to the DMF and/or application. Please note that such amendments may have user fee goal date implications and will be assessed accordingly.

8. ARB API producers evaluate each process for the potential to form a nitrosamine impurity and if at risk, make changes necessary to prevent nitrosamine formation. If the process cannot be changed to prevent nitrosamine formation, FDA will permit use of a robust purging/elimination step(s) provided that it includes an appropriately sensitive test to verify that the resulting intermediate or API does not contain a detectable nitrosamine impurity. See existing FDA guidance in ICH Q7 *Good Manufacturing Practice Guidance for Active Pharmaceutical Ingredients*, ICH Q11 *Development and Manufacture of Drug Substances*, and ICH M7.

9. Batch testing to verify no detectable nitrosamine in the API should continue unless the API producer has demonstrated their process is not at risk for producing detectable nitrosamine in accordance with guidance (see, e.g., ICH Q7). This includes demonstrating that:
   - starting materials, including vendor-supplied intermediates, have no detectable nitrosamines or such amounts can be purged such that the API contains no detectable amounts of nitrosamines, and

- raw materials used in the process, including recovered solvents and catalysts, contain no detectable amounts nitrosamines.

10. ARB API producers should voluntarily report the finding of a nitrosamine impurity to FDA in a Field Alert Report even if the contaminated material was not used in API processing. FDA will review the reports to determine if other API producers are unknowingly at risk to nitrosamine contamination and notify accordingly.

11. ARB API producers report to FDA a summary of the testing performed, as requested above, for the presence of any nitrosamine impurities in batches distributed in the US, whether directly as an API or after incorporation into a DP for the US, that are within their labeled expiration or retest-by date, even if recalled. We request a table be submitted to each DMF, if not already provided in previous correspondence to the DMF, with the following information for each batch number that is sampled and tested:

12. • API name
   - date of manufacture
   - labeled expiration or retest-by date
   - name of test method
   - amount and type of nitrosamine detected, if any, or "none detected."

Data should be submitted to the Drug Master File as "Quality/Controls;" the words "nitrosamine-related" should be prominently displayed on the cover letter.

13. ARB API producers report to each ARB DMF information about each independent facility that recovers materials used in ARB production for the past two years. We request the following information about such facilities: business name; address; name of recovered material; and, the month and year the recovery facility has been performing recovery operations for the ARB.

FDA will, to the extent possible, expedite review of amendments and supplements for manufacturing changes required to eliminate or limit a nitrosamine impurity, or when needed to prevent or mitigate a drug shortage.

For the testing requested in this letter, the detectable amount of nitrosamine impurity should be based on one of the following:

1. The limit of detection established in one of FDA's published methods.
2. A method published by another regulatory agency that is equivalent to FDA's method(s).
3. Any appropriately developed and validated method capable of an LOD and Limit of Quantitation (LOQ) equivalent to a method published by FDA.

The interim acceptable limits and FDA published methods, as well as other FDA information on this issue, are available at https://www.fda.gov/Drugs/DrugSafety/ucm613916.htm.

FDA published methods have been validated to detect and quantify NDMA and NDEA in all ARB APIs and some ARB DP formulations. FDA may update existing methods or post new methods once validated for use in detecting other nitrosamines in DPs and APIs. FDA may also update its published methods to improve their limits of detection and/or quantitation; if updated, FDA expects that manufacturers will update their methods to achieve comparable limits and apply the new LOD, if any, in making decisions about batch suitability.

You should share this letter with your suppliers (e.g., solvent recovery vendors and starting material suppliers) and contract manufacturers.

Appendix VIII: Master Qualification Plan (MQP) – Template

Master Qualification Plan (MQP) for _____ describes the extent, the activities and the responsibilities with the qualification and validation of rooms, equipment, Piping loops, and computerized system frameworks of the new, or upgraded _____.

Engineering			
Name	**Department**	**Date**	**Signature**
	Engineering		

Review / Examination			
Name	**Department**	**Date**	**Signature**
	Engineering		
	Validation		
	QA		
Approval			
Name	**Department**	**Date**	**Signature**
	Engineering		
	Production		
	Maintenance		
	Validation		
	Compliance		

1. GENERAL INFORMATION

1.1 INTRODUCTION

Facility planned the establishment of new, or upgrade _____ to produce _____. The equipment is manufactured by _____, USA, OTHER and is composed of _____. Controls are designed to perform 100% (In Process Control (IPC), or as specified.

The documentation requirements for this system involve the development of MQP, HLRA, URS, TM, MDS, HDS, SDS, FDS, DR&DQ, FAT, C&Q, SAT, IQ, OQ, PQ, and Operations Manual and Maintenance Manual.

1.2 OBJECTIVES

This master plan defines the fundamental approach and responsibilities for the qualification activities. With the qualification to be guaranteed that the system associated in each case with control systems meets the cGMP requirements, which applies regulatory requirements and the conditions corresponding to the technology being suitable for the intended purposes. Thus, it is to be guaranteed that in this application the products manufactured or used by this system exhibit the necessary quality attributes.

1.3 SCOPE

This master plan is valid for the qualification of the project: new or upgraded _____. This covers project engineering lead and other team members to ensure execution of applicable documentation.

2. DEFINITIONS AND ABBREVIATIONS

Term	Description
URS	User Requirements Specification
MDS	Mechanical Design Specification
cGMP	current Good Manufacturing Practice
CSV	Computerized system Validation
DQ	Design Qualification
FAT	Factory Acceptance test
SDS	Software Design Specification
FDA	Food and Drug Administration (USA)
FDS	Functional Design Specification
HDS	Hardware design Specification
IPC	In Process control
IQ	Installation Qualification documented proof that the installed and/or modified plant or system with the authorized design, which agrees on manufacturer data and operator requirements.
MQP	Master Qualification plan
DR	Design Review
OQ	Operational Qualification documented proof that the installed and/or modified plant or system as intended in accordance with the given work areas functions.
FDS	Functional Design Specifications
PQ	Performance Qualification documented proof that the plant and its connected auxiliary systems effectively and reproducibly, perform based on authorized processes and specifications work.

(Continued)

Term	Description
C&Q	Commissioning & Qualifications
HLRA	High-Level Risk analysis
P&ID	Piping and instrumentation Diagram
SAT	Site Acceptance test
TM	Traceability matrix

3. BASES

3.1 DEFAULTS

This master plan follows guidelines of Process validation, equipment Qualification, Classified room qualification, and Computerized system Validation.

The following Client internal multi-site SOPs and standard regulations continue to be valid with the definition of the individual qualification activities:

- Change control Policy
- Deviation SOP
- Equipment Qualification SOP
- Computerized system Validation SOP
- Preventive Maintenance SOP
- Calibration of instruments SOP
- Other?

3.2 ACTION

The qualification of the new or upgraded _____ is based on the project URS, the specifications (MDS, HDS, SDS – FDS) and the associated risk analysis. In this framework, hig- level GMP risk analysis (HLRA), the qualification-requiring systems of the new, or upgraded _____ are identified. For these systems the general qualification action in this MQP is specified. Details for qualification are defined in the respective qualification plans.

All qualification steps of the _____ and infrastructure can be implemented. For the respective partial qualification after successfully accomplishing PQ and line release, an interim report is written. A summary of the interim reports takes place in the final qualification report.

3.2.1 URS

As a condition for the qualification, each system must be shown in an authorized URS. The systems and their components can be illustrated in one or more URS. The authorized URS is the basis of all qualification activities.

The URS supplies an abridged, concise description, which describes the basic characteristics and function requirements of the system. These elements are illustrated and checked during the qualification in different minutes. Under the URS the individual requirements are already classified whether they are GMP and/or EH&S relevant. All remaining requirements are GEP and/or EH&S relevant.

3.2.2 Traceability matrix

The purpose of the Traceability matrix is to bridge between URS and GMP risks, and the test plans of all GMP-relevant requirements (FDS).

GAMP system collection and evaluation (CSV category)

The GXP of relevant computerized systems of the systems is captured in the CSV High-level Risk analyses. Typical computerized systems are assembled from different components of the different GAMP categories. This should be illustrated with the CSV system evaluation.

3.2.3 Risk analysis

The GMP-relevant systems are qualification requirements. The extent and the depth of the qualification depend on the kind and the criticality of the system and/or the component. The criticality of the individual systems and/or components is evaluated within risk analysis.

GMP risk analysis is accomplished in relevant systems and components to be evaluated and measures for minimizing the risk specified. The measures identified in the risk analysis are accomplished and documented in the qualification phases. The definition of the risks of adjacent systems and utilities are regarded separately. For an explicit view and evaluation of the EH&S risks an EH&S risk analysis is provided.

3.2.4 Specification (equipment/process requirement specifications)

Specifications like MDS, SDS, HDS, and FDS, where applicable, the user demands for the new system, converted by the supplier, document. These specifications are provided in principle by the supplier. In each case they must be checked by the technical and quality departments of Client and be approved in the context of the DQ. Associated supplements (e.g. designs, sketches or detailed descriptions) are attached to these specifications with appropriate reference.

3.2.5 Training

All users should have completed a documented training before release and use of the system.

3.2.6 Design Qualification

The DQ isa combination of DQ-plan-relevant documents. It contains the examination of the respective URS specified documents and plans and general agreement with the determining guidelines and regulations. In addition, the supplier specified specifications are adjusted against the requirements from the URS. In the DQ the plans are approved. These plans are used in the following qualification phases as inspection procedures.

3.2.7   FAT and SAT

3.2.7.1   FAT and SAT for _____ (mechanical part)

In the context of the FAT the systems and components (equipment, instrumentation, documentation, etc.) are checked with the supplier for completeness and for functions and installation. Switchgear cabinet controls with I/O test are accomplished. In addition, the material and calibration certificates as well as the technical documentation for completeness are examined. For control programs in the FAT, correct and complete programming is usually tested. Protocol/report governs these activities.

In the context of an SAT, examinations are accomplished, which were not yet possible at the time of the FAT, or which is repeated due to transport of the supplier equipment on the site. Also pending items from the FAT are usually examined in the context of the SAT for completion. The FAT and SAT documents are usually provided by the supplier, or written protocol by Owner to achieve specific testing requirements. They must be checked and approved, however, before execution by the technical and quality departments. Upon completion of SAT, commercial aspects such as warranties on equipment or labor and materials coverages are initiated. This is important as sometimes equipment installations are delayed due to construction schedule constraints, but revision of contract timings must be observed carefully.

3.2.7.2   FAT and SAT for controls (CSV)

The software for processing units is tested later usually first in a test environment with the supplier and at the place of assembly. The control programs are examined here for correct and complete programming.

The FAT and SAT documents are usually provided by the supplier. They must be checked and approved however before execution by the technical and quality departments of Client.

3.2.8   Installation Qualification

The IQ contains the verification of the correct and complete installation in accordance with the planning documents.

The IQ covers the examination of:

- Correct assembly with consideration of the "Good Engineering Practice"
- Adherence to the regulations of cGMP and ISO 9001
- Installation in accordance with the valid local regulations
- Marking and inscription of the components
- FMECA must be completed to assess risks at granular level, including all process connections

A goal of the IQ is to furnish the documented proof that in accordance with the accomplished installations in the DQ approved plans and specifications as well as the valid regulations. The IQ checks documented controls and tests, those of components, during the production in the work (FAT) and during the assembly (SAT), but particularly after final assembly as preparation for start-up.

IQ-plans provided by external suppliers can be used for the tests specified in the work instructions specified above and for further qualification activities, which are not yet taken off by the work instructions specified above, identified in the risk analysis.

A condition for the use of these plans is that the activities specified in the IQ-plans at least cover the measures identified in the risk analysis adequately, which work instructions correspond, and that the plans are approved by Client technical and quality departments.

3.2.9 Operational Qualification

The OQ contains the documented verification of the fact that the assigned equipment in accordance with which defaults within the defined instrument range operates and which functions to specifications accordingly. During the OQ, both normal operation and the behavior must be examined with alarms and disturbances.

OQ plans provided by external suppliers can be used for the examinations specified in the work instructions specified above and for further qualification activities, which are not yet taken off by the work instructions specified above, identified in the risk analysis.

A condition for the use of these plans is that the activities specified in the OQ plans at least cover the measures identified in the risk analysis adequately, which work instructions correspond, and that the plans are reviewed by Client technical and quality departments.

3.2.10 Performance Qualification

The goal of the PQ is to furnish the documented proof that in accordance with PQ plan- defined critical parameters within the given specifications (work area, result, quality), reliably and reproducibly of work and the product /equipment/process corresponds to the demanded quality. The PQ contains besides the proof that the cooperation of the individual systems covers the expected production department.

The risk analysis identified the performance qualification activities by owner, or by external suppliers provided PQ plans are used. A condition for the use of these plans is based on the previous examination and approval of the plans by the technical and quality departments. Process flow diagram and batch records are to be used in support of the test procedure.

PQ tests can be accomplished with substitute or current production material, depending upon the kind of system.

4. ORGANIZATION AND RESPONSIBILITIES

4.1 Defaults

Project organization represents the functional structure of assigned persons. Personnel changes within functions do not require update of this MQP.

4.2 Responsibilities

The following table gives an overview of the involved departments and competencies for production, examination, and approval of the subordinate documents. Details of execution of DQ, IQ, OQ, and PQ including CSV are specified in the appropriate qualification and phase plans

	ENG	PROD	MAINT	QA	EH&S	IT	External company
MQP (master Qualification plan)	x	x	x	X	x	x	
Project URS	x	x	x	X	x	x	x
High level GMP risk analysis	x	x		X			
MQR (master Qualification report)	x	x		x		x	

4.3 EXTERNAL EMPLOYEES

The project is implemented in cooperation with external enterprises of different technical suppliers, which are qualified.

It is to be guaranteed that execution takes place from qualifying measures via external employees in conformity to the procedures of Client Site and if necessary authorized qualification plans of the respective company.

Documented training for the external employees for execution and documentation of qualification activities are on record. An admission of such an activity must only take place after conclusion of the documented training. Details are defined in the respective qualification plans.

5. QUALIFICATION EXTENT AND DOCUMENTATION

5.1 QUALIFICATION EXTENT

The qualification extent for the individual systems and components is specified in the respective qualification plan.

CSV system	Doc. Number	Title	Extent
		process control system / co-ordination control	
		Qualification/validation plan	
		Qualification/validation plan	
		Qualification/validation plan	
		Qualification/validation plan	

5.2 DEMARCATION TO OTHER SYSTEMS

5.2.1 Demarcation to the SCADA system

The SCADA system is a validated computerized system which serves the recording and archiving that GMP-relevant process data at the location, and Site. The defined data are made available for a unidirectional data transfer over an interface to SCADA in data components.

The associated servers will receive the data which can be archived are fetched via gateway. The qualification/validation of the new controls has the documented proof of the physical presence of the interface to supply the specified data parameters to data transfer. Likewise, the documented proof is to furnish that the data transfer and the data integrity of the data transfer of the defined data up to the interface are ensured.

5.2.2   Demarcation to the system SCADA

SCADA is a validated computerized system which serves as facility management system.

The HVAC systems of classified rooms are to be steered over SCADA, the extension of the existing system corresponds to validation of recorded data

5.2.3   Demarcation to Active directory and IT-network

Active directory is a validated system to the user authentication with multiple levels of access. The user authentication for all production recipes is included in the process control system. In the qualification plans the qualification tests for each qualification phase are defined.

5.2.4   Acceptance criteria

The acceptance criteria are per test and main component or system, if not generally valid to formulate specifically. Only measurable and/or controllable criteria are to be defined. A test is successfully final only if all defined criteria are fulfilled and no serious pending concerns in quality and security.

5.2.5   Qualification documentation

The qualification documentation for GMP-relevant items of equipment must be developed such that the necessary traceability is ensured.

Usually, client internal qualification documents are used according to the work instructions. However, check whether the existing, authorized forms for standards are enough in the risk analysis identified measures or whether project-specific adjustments and/or additional qualification documents are necessary, provided qualification documents and/or qualification documents of the supplier can be used, if they correspond to the requirements of the valid internal regulations and were checked and approved by the specialized division and the responsible quality assurance representative.

If the provided qualification documents and/or the qualification documents of the supplier take into consideration all of the risk analysis identified measures, then work instructions mentioned upward without an additional execution can be used.

Records of qualified personnel or qualified external employees comply with GMP and with a document-quality writer (no pencil) to fill out test results, whether the acceptance criteria are fulfilled or not to document in the respective test log.

On the inspection device the checkpoints must be checked off and be dated at the conclusion of the inspection. The qualification dossier must be arranged compellingly from the original diagnostic data sheets (test reports). In addition, calibration records, material certificates, technical documentation, and signatures on the test results must be included.

The filled-out Q-documents are archived by the quality assurance department.

6. CHANGE CONTROL/DEVIATIONS

Project Change Control is conducted in accordance with quality systems. All changes must be formally judged and approved. The quality department decides whether for a change a separate request for modification must be provided. For cGMP-relevant changes a request for modification is mandatory.

Change control refers particularly to the authorized documents (specifications, loads/product requirement specifications etc.) in the design Qualification. The revised documents must be clearly recognizable, i.e. a new version number or an index with new date must be assigned. In addition, it must be evident which version was replaced, and which changes were made.

Deviations in the project are documented with the qualification in accordance with quality systems procedures. Additionally, the deviations in the discrepancy list of the test log are combined and submitted to the quality department for evaluation. All deviations are registered in the project spreadsheet during the qualification. A system is regarded only as completely qualified if all substantial and critical deviations are closed.

7. RELEASE OF THE AREAS AND LINES

After conclusion of the qualification of a system a final report is provided and the qualified system for the intended purposes is released. Changes and adjustments on systems are approved with conclusion of the respective CRs and the respective qualification reports.

If no substantial or critical pending issue of the DQ/IQ/OQ/CSV and PQ is present and if all plan reports were approved, the qualified systems can be released to the inclusive control systems before production and approval of the final report.

After conclusion of the qualification of all systems the master qualification report, which formally concludes the qualification of the new line, is provided.

8. DEVELOPMENT

Version	Date	Change/Reasons for the revision
01		
02		
03		
04		

Appendix IX: High-Level Risk Analysis – Template

High-level GMP Risk Analysis serves to meet the requirements for new or upgraded facility. This contrasts a fundamental distinction between GMP- and GEP-relevant system.

Client planned the establishment of new or upgraded site or system to produce pharmaceutical drugs. The installation phase is planned to install systems in classified rooms under ISO 5, 6, 7, or 8 clean room conditions

This high-level GMP risk analysis serves to contrast fundamental differences between GMP- and GEP-relevant system. It moves from there on a high system level and does not replace detailed GMP risk analysis, which is based on the different URS for individual system components and parameters, and Impact Assessment. Following an FMEA, which contains the qualifying measures that can be accomplished, these measures are then specified.

Term	Explanation
WFI	Water For Injection
STS	Solutions Transfer System
Mix	Mixing Building
RTU	Ready-to-use
BMS	Building Management System
CIP	Cleaning in Place
FMEA	Failure Mode and Effects Analysis
GEP	Good Engineering Practice
GMP	Good Manufacturing Practice
RABS	Restricted Access Barrier System
LAF	Laminar air flow
HLRA	High-level risk analysis
IT	Information Technology
PLC	Programmable Logic Controllers (Process control system)
SCADA	Supervisory Control And Data Acquisition (Computer System)
RA	Risk Analysis
EH&S	Environment, Health & Safety
URS	User Requirement Specification

SYSTEM DESCRIPTION

Systems have mechanical features that are managed by control systems and electrical cabinets that are composed of high voltage power supply and low voltage for control systems. The RTU system produces and distributes process gases, liquids, and powders through transmission piping, valves and instrumentation. Compounded Liquids are then transported to the filling zone in position for filling specific primary packaging components. Packages are then transferred to terminal sterilization, printing/coding, and packaging stations for release.

Insert picture_____

Illustration: Photo representation _____ --

Building/area/ layout _____

Insert DWG: Room layout / Equipment layout / arrangement

Procedure to produce _____ Risk Analysis

The HLRA targets an Impact assessment. All systems are to be listed and evaluated regarding their influence, in order to specify the qualification obligation for the individual systems of apparatuses.

The following allocation of "impacts" to the aspects of the 1–6 points below is to be used:

1) Does the system have direct product contact (inclusive raw materials)?
   => Direct impact system
2) Does the system manufacture a product?
   => Direct impact system
3) Is the system used for the cleaning?
   => Indirect impact system
4) Conditioned – the system influence product quality (e.g.: Nitrogen blanketing, material storage condition, e.g. temperature)?
   => Indirect impact system
5) Does the system produce quality-relevant data (e.g. quality relevant alarm)?
   => Indirect impact system
6) Is the system a process control system (e.g. PLC) or does it contain a process control which has influence on product quality?
   => Indirect impact system

The qualifying measures, for systems with the status "Direct impact system" and "Indirect impact system", are specified later in detailed GMP risk analyses.

Execution: Impact assessment

System component	Product contact	Production product	Cleaning	Product quality	Data	Process control system	Direct impact system	Indirect impact system	NO impact system	Remarks
Building/building engineering										
Building entrance								X		Entrance protection must be present
Division in zones				X				X		-
Ventilation systems									X	-
Space conditions									X	EH&S
Space equipment									X	-
Ventilation systems zone D inclusive air-locks				X				X		-
Space conditions zone D inclusive air-locks				X				X		-
Space equipment zone D inclusive air-locks			X					X		Cleaning ability, surfaces
Video monitoring of production areas								X		At least for handling of final stages

(Continued)

System component	Product contact	Production product	Cleaning	Product quality	Data	Process control system	Direct impact system	Indirect impact system	NO impact system	Remarks
Tank compartments									X	No storage
Technology area									X	-
Storages, tool-storage rooms									X	-
Office									X	-
IT-infrastructure					X			X		Extension existing net
BMS					X			X		Extension existing system
Utilities										
Steam									X	
Compressed air								X		Room ____ only control air
Source exhaust air				X				X		Final constant filters
Pure water	X						X			Plant (production) and loop

(Continued)

System component	Product contact	Production product	Cleaning	Product quality	Data	Process control system	Direct impact system	Indirect impact system	NO impact system	Remarks
Clean Air	x						X			
Nitrogen				X			X			
Process engineering										
Product lines and stations	X	X					X		-	
CIP systems			X					X	-	
Vacuum - pumps				X				X	-	

(Continued)

System component	Product contact	Production product	Cleaning	Product quality	Data	Process control system	Direct impact system	Indirect impact system	NO impact system	Remarks

Change history

Version	Date	Reason of the revision	originator
01			

Appendix X: Cleaning Validation

Cleaning Validation Master Plan (CVMP):

- Detail the strategies and requirements for the Cleaning Validation (CV) protocols for the product process equipment including powder transfer systems, mix tanks, solution transmission systems (STS), fill machines, and parts washers. This write-up does cover sanitization process or requirements for WFI.
- Describe processes to be completed prior to execution of the cleaning validation activities.
- Outline requirements pertaining to cleaning agents, cleaning procedures' acceptance criteria, tasks and expectations during the cleaning and sanitization validations established. These requirements will be based on policies and procedures, applicable country regulations and guidelines along with acceptable industry practices for cleaning validation.
- Cleaning development and validation activities will be executed in accordance with this CVMP.
- Ensure that all manufacturing equipment is operating under a comprehensive cleaning program and cleaning validation system.

This CVMP details CV requirements for Stages 1 and 2 of the CV process lifecycle, including Cleaning Process Design and Development (Stage 1) and Cleaning Validation (Stage 2). Continuous Process Verification (CPV) (Stage 3) will be established following completion of validation.

SCOPE

This document applies to Site facility and is intended to outline and detail a comprehensive strategic plan for the Cleaning Validation program at this facility for all solution (product) contact formulation and filling manufacturing equipment systems.

The scope of this CVMP is all the product contact equipment and parts used to manufacture pharmaceutical products at the Site Facility. The cleaning systems indicated in the following table will be used to develop cleaning processes of direct and indirect product contact surfaces for process and dispensing equipment/systems identified. System boundaries of each system will be defined in individual protocols.

Cleaning Systems

Cleaning System
CIP Skids (Mobile and Stationary)
Roto Jet (Rotary Spray Device)
Power Cleaning Vat (Agitated Immersion)
Pipe Vat (Static immersion)
Sonicator (Agitated, heated immersion)
Sani-Matic small parts washer

Process Equipment/Systems

Process Equipment/Systems
Mix Tanks
Mixers
Surge Tanks
Holding Tanks
Pumps
Filter Housings
Solution Transmission System (STS)
Filling Nozzles
Ancillary Parts

DEFINITIONS AND ABBREVIATIONS

Acceptable Daily Intake (ADI) – A measure of the amount of a specific substance that can be consumed on a daily basis over a lifetime without any appreciable health risk.

Active Pharmaceutical Ingredient (API) – Any substance or mixture of substances intended to be used in the manufacture of a drug product that when used as such becomes an active ingredient of the said drug product. Such substances are intended to furnish pharmacological activity or other direct effects in the diagnosis, cure, mitigation, treatment, or prevention of disease or to affect the structure and function of the body.

Acceptable Residue Level (ARL) – The highest amount or concentration of a specific material that is permitted on the equipment surface prior to a subsequent production operation using the identical equipment. The ARL is calculated from the MAC.

Baseline Visual Inspection – Documentation of permanent stains and surface imperfections prior to conducting the validation study to avoid potential false positive failures.

Challenge Soil (Mock Soil) – A formulated or process soil that represents a worst-case cleaning challenge for the system as used to support cleaning procedure or CIP cycle development.

Clean Hold Time (CHT) – Maximum time allowed from the end of the cleaning process until the beginning of the use of the cleaned equipment for manufacturing.

Clean In Place (CIP) – A system or process of cleaning that involves cleaning equipment without disassembly of the equipment. CIP systems typically include a CIP unit (skid) that is composed of storage tank(s), a heat exchanger, chemical feed equipment, circulation pump, process control devices, and other instrumentation, one or more spray devices and associated piping.

Clean Out of Place (COP) – A system or process of cleaning equipment items by removing them from their operational area and taking them to a designated cleaning station for cleaning. It requires dismantling an apparatus, washing it in a central washing area using an automated or manual process.

Cleaning Agent – A solution or solvent used in the cleaning process, which promotes removal of residual product and other contaminants from the equipment surfaces. Examples of cleaning agents are water, organic solvent, commodity chemical diluted in water, and formulated detergent diluted in water.

Cleaning Validation – Documented evidence with a high degree of assurance that a cleaning process will result in products meeting their predetermined quality attributes throughout its life cycle.

Continuous Process Verification (CPV) – Process that provides documented evidence that a cleaning procedure remains in a state of control after cleaning validation is performed. The frequency of monitoring studies is determined through a risk assessment.

Critical Cleaning Parameter (CCP) – Cleaning Parameters, such as TACT, which if any of the parameters are not met, the resulting process may not meet predetermined acceptance criteria.

Cycle Development – Characterization of soil and equipment conditions to develop a consistent and effective cleaning methodology for clean-in-place processes.

Critical Site (Hot Spot) – A difficult to clean location in equipment, which if contaminated may lead to the contaminant being distributed in the next batch of product.

Dedicated Equipment – Equipment that is only used to manufacture a single drug product.

Dirty Hold Time (DHT) – Maximum time allowed between the end of manufacturing and the beginning of the cleaning process.

Manual Cleaning – Cleaning procedures

Maximum Allowable Carryover (MAC or MACO) – The highest amount or concentration of a specific material that is permitted to be carried over into a subsequent batch in the same equipment.

Mock Soiling – Soiling equipment surfaces that would represent production conditions.

Product Contact – Equipment that contacts product through normal production.

Recovery Factor (RF) – A factor that converts the amount of residue detected in a sample to 100% by accounting for the amount of residue that cannot be recovered during sampling.

Rinsate – (also referred to as "rinse water"). A representative aliquot of effluent rinse water that is used to determine the adequacy of cleaning and rinsing procedures.

Swab – An absorbent pad or piece of material used for cleaning validation sampling operations.

Visually Clean – The surface of the equipment shows no visible residue from product, cleaning agent, or other components.

Worst Case Scenario – The worst-case scenario is determined based on process soils, sampling, and/or cleaning conditions that would challenge the cleaning process to meet predetermined acceptance criteria.

ABBREVIATION	DEFINITION
ADI	Acceptable Daily Intake
API	Active Pharmaceutical Ingredient
ARL	Acceptable Residue Level
CA	Cleaning Assessment
CAPA	Corrective and Preventive Action
C&Q	Commissioning and Qualification
CCP	Critical Cleaning Parameter
CHT	Clean Hold Time
CIP	Clean-in-Place
COP	Clean-Out of-Place
CV	Cleaning Validation
CVMP	Cleaning Validation Master Plan
CQA	Critical Quality Attributes
DHT	Dirty Hold Time
DW	Distilled Water
EHS	Environmental Health & Safety
GEP	Good Engineering Practices
GMP	Good Manufacturing Practices
GRAS	Generally Recognized As Safe
IOQ	Installation and Operational Qualification
LOD	Limit of Detection
LOQ	Limit of Quantitation
MAC/MACO	Maximum Allowable Carryover
MAR	Maximum Acceptable Residue
MOC	Materials of Construction
PACE	Process and Cleaner Evaluation
P&ID	Piping and Instrumentation Diagram
PDE	Permissible Daily Exposure
PFD	Process Flow Diagram
PPQ	Process Performance Qualification
QA	Quality Assurance
QC	Quality Control
RPS	Residue Per Sample
SOP	Standard Operating Procedure
SCS	Standard Cleaning Solution
STS	Solution Transmission System
TACT	Time, Action, Concentration, Temperature
WFI	Water for Injection

GENERAL REQUIREMENTS

Requirements for cleaning validation are based on direct/indirect or non-product contact surfaces. All equipment/systems within the scope of this CVMP are considered multiple use systems and will require demonstration of the following:

- Visually Clean
  - Equipment is visually clean (absence of product and cleaning agent residue and absence of pooling of water) upon inspection under adequate lighting conditions by a trained/qualified inspector.
  - Tools such as mirrors with extended poles and/or borescopes may be used to inspect areas that are not readily accessible for visual inspection with naked eye.
- API Removal
  - Reduction of potential API residual is challenged and demonstrated to be within acceptable pre-determined levels
- Major Ingredient Removal
  - Reduction of worst-case ingredient is challenged and demonstrated to be within acceptable pre-determined levels.
- Cleaning Agent Removal
  - Reduction of cleaning agent residual is challenged and demonstrated to be within acceptable pre-determined levels. Cleaning agents are not part of the process and their reduction to within acceptable levels must be demonstrated. A toxicological assessment of the cleaning agent to determine acceptable residue levels is recommended.
- Bioburden Control
  - Reduction of potential microbial residual to pre-determined levels is demonstrated during clean and dirty equipment hold time validations.
- Endotoxin Control
  - Reduction of potential endotoxin to pre-determined levels is demonstrated.

CLEANING VALIDATION PLAN

The Site cleaning validation program is intended to establish documented evidence that provides a high degree of assurance that specific cleaning processes consistently clean manufacturing equipment to predetermined acceptable levels.

The CV process life cycle exists within a matrix of activities that form the lifecycle for commercial production. This life cycle begins during the initial cleaning activities performed during product and process development (Stage 1) and proceeds through validation (Stage 2) and continuous process verification (Stage 3).

As part of Stage 1 and Stage 2, the following activities will be performed:

All existing cleaning procedures will be evaluated to ensure that they are adequately validated. All equipment will be evaluated to ensure that they are qualified and that the cleaning procedures for each item have been satisfactorily validated. Cleaning limits for products and detergents will be assessed and the worst-case product will be identified and justified. New cleaning procedures will be developed for all systems that lack adequate cleaning validation.

CLEANING PROCESS DEVELOPMENT (STAGE 1)

The following activities are considered during Stage 1 Cleaning process development:

- Cleaning Development Studies are required for manual, semi-automated, and automated cleaning processes.
- Define system to be cleaned, including equipment, manufactured products, and potential contaminants.
- Design and develop cleaning procedures, including control of Critical Cleaning Parameters (CCP).
- Design overall cleaning validation approach, including residues to be measured, sampling methods and sampling locations, and determination of worst-case products. Use of placebo instead of product must be justified, and any mock soiling should be described and justified.
- Determine Dirty Hold Times and/or Clean Hold Times.
- Define acceptance limits (calculate MAC).
- Determine the type, concentration, and other usage/action parameters of suitable cleaning agents.
- Confirm compatibility of system MOC and cleaning agents.
- Develop and validate analytical methods to be used. Minimally, the LOD and LOQ are determined prior to using the analytical methods to test development samples.
- Develop and validate or qualify microbial methods to be used in development studies.
- Perform recovery studies on all major product contact surface materials and all surfaces that will be swabbed.
- Generate cleaning procedure or cleaning cycle development report(s) to capture all pertinent information, data collected, and rationales.

EQUIPMENT/SYSTEM CLEANING REQUIREMENTS

Systems identified within this plan include single equipment systems and equipment groups (i.e., process train/flow) required to perform a given manufacturing activity.

Cleaning procedures for any new equipment/systems will be developed and validated to demonstrate that it meets the current standards. New equipment/systems will be evaluated against the worst-case product matrix. New cleaning validations will be performed using the worst-case equipment and the worst-case product(s) to be manufactured on the equipment/system.

Equipment/systems outlined in this CVMP are generally intended for multiuse purposes, where they are not dedicated, and so it must be demonstrated that cleaning procedures are robust and effective in reducing residues within acceptable limits and thus minimize the potential for cross-contamination.

SURFACE AREA CALCULATIONS

Surface areas for each equipment will be calculated following good engineering practices (GEP). Drawings of individual equipment with the product contact surface areas will be kept in a controlled document. The total shared surface area will be used to calculate the residue per sample (RPS).

LIMITS ESTABLISHMENT

Limits for contaminants will be established, based on scientific knowledge and what is practical, achievable, and verifiable. For protocols, product contact surface residue acceptance criteria shall be established for chemical (drug product and cleaning agent) contaminants and microbiological residues (bioburden and endotoxin). In addition, accessible equipment surface shall be verified visually clean.

Drug Product (API or major ingredient) and cleaning agent residue limits criteria are established using a calculated limit that is based on potential carry-over into a product following the subject batch. Limits are established based on surface areas of the manufacturing shared equipment train and toxicological reports for maximum daily dose of residue.

Limits for microbiological contaminants will be based on the quality of the final rinse water and the release criteria for the final products.

CRITICAL CLEANING PARAMETERS

Critical cleaning parameters; time, action, chemistry/concentration, temperature (TACT) will be established during development of cleaning cycles and associated cleaning procedures for all pre-wash, wash, rinse, and drying steps:

Time

Time is the duration of a specific cleaning or drying step. Time may be established by fixed value, bracketed value, or minimum duration.

Action

Action is "how" a specific cleaning step is performed.

Manual cleaning processes typically describe 'Action' by (1) use of cleaning implement (tool) and (2) technique (e.g. scrub back-forth using overlapping strokes, etc.).

Automated cleaning processes typically describe 'Action' by (1) use of cleaning system and (2) technique (e.g., spray-device impingement, cascade flow, etc.).

Action may be established by fixed value, bracketed value, or minimum duration.

Chemistry/Concentration

Chemistry/Concentration is "what" a specific cleaning step is cleaned or rinsed with. CIP100 /200 (Steris), or Minncare is used at concentration of 0.75% for the mix tanks and certain STS components such as Robotic Fill Machine (RFM).

Chemistry/Concentration will be established by a fixed target value. Once an optimum value is determined, this value will not be challenged during validation as increasing the value may require more rinse water and reducing the concentration may not accomplish appropriate cleanliness.

Temperature

Temperature is the required temperature for a specific cleaning step, including wash, rinse, and drying steps.

Temperature may be established by a fixed value or a range. As a rule of thumb, cleaning action may double for every increment of temperature by 10°C. However, ambient or lukewarm temperature is suitable for prewash if the product or excipient reacts undesirably with hot water. Hot temperature conditions are preferred for the final rinse water to enhance drying and to reduce microbial proliferation.

CLEANING VALIDATION MATRIX APPROACH

The strategy for a validation study and rationale for the study are determined on a case-by-case basis according to the guidelines documented herein and documented in the associated protocol.

Cleaning Validation Matrix

	Matrices to consider for Cleaning Validation Studies
Family Grouping	1. Family grouping is a strategy whereby manufactured products and/or equipment are considered together, and a formal protocol is performed on a representative from the group.
	2. The representative from the group is usually the worst case among the products or equipment in a group.
	3. Product and equipment must be cleaned by the same cleaning procedure to be considered for grouping.
Equipment Grouping	1. Identical equipment are equivalent in all respects such as surface finish, product contact surface area, materials of construction (MOC), disassembly/cleanability, etc. Equivalency must be demonstrated based on the IQ/OQ of the equipment. Any one of the equipment can represent that group. However, if it is used for different processes, the equipment used for the most difficult process should be selected for CV studies.
	2. Equipment equivalent in all respects except for the size can be grouped together in one family for cleaning validation purposes. CV studies may be performed on the smallest and the largest size which is considered as the worst case.
	3. Equipment that are equivalent in most areas except for some minor aspects may also be grouped based on a risk analysis. The worst-case equipment based on a risk analysis may represent the group for CV studies. The worst-case equipment may undergo a full validation and at least one run may be required for other members of that group.
	4. Length of downstream piping, nozzles per fill machine, and number of filling machines are also factors that need to be considered for grouping strategy.

(Continued)

Matrices to consider for Cleaning Validation Studies	
Product Grouping	1. Product grouping is based on the premise that if the cleaning process is effective for the worst-case product, then it would also be effective of the other easier to clean products in the group.
	2. Worst-case products are selected based on the solubility, toxicity, concentration, stability, etc.
	3. A matrix of product versus equipment will be prepared to determine the matrix approach to cleaning validation.
	4. One product from the group, or sometimes more than one product based on a risk analysis, may represent the group for CV studies.
New Equipment	1. New equipment will be compared against other "like" equipment on site and will be grouped accordingly based on risk assessment and grouping strategy.
	2. For equipment that cannot be grouped with existing equipment, three consecutive successful validation runs are required.

WORST-CASE PRODUCT/EQUIPMENT

Grouping strategy will be used which involves grouping similar products and function and determining a "worst-case" representative of the group. The basic premise is that if the cleaning process is effective for the worst-case product or equipment, then it would be effective for the other easier to clean products or equipment in the group.

The following criteria is used for the selection of the worst-case product in a group:

- Potency – expressed as daily dose (in mg or mcg). The lowest dose will be used when the product is given in a range of doses since this would represent the worst case (the lower the dose, the higher the potency of any given product).
- Toxicity – expressed as LD_{50}. The lowest LD_{50} value will be used since that represents the most toxic product. This is used as the criteria to determine the worst-case ingredient in a cleaning agent.
- Solubility – expressed quantitatively as e.g., mg/ml. The least soluble product in the cleaning medium will be used as the worst-case product for cleaning validation purposes.
- Cleanability – this is a very subjective characteristic of a product and is empirically determined since it is not predictable from scientific data. Operator experiences or laboratory studies may be used to rate the difficulty of cleaning.

CLEANING PROCESS VALIDATION (STAGE 2)

Execution of cleaning validations is performed to confirm the cleaning process is consistently effective at reducing residuals to a predetermined acceptance level. This methodology applies to all systems discussed within this plan, including manual, semi-automated, and automated cleaning systems. The CV testing methodology must test the cleaning of equipment under actual process conditions.

The following activities are performed during Stage 2 Cleaning Validation:

- Prepare and approve cleaning validation protocol.
- Execute cleaning validation protocol by trained personnel.
- Recovery studies must have been completed for all MOCs that will be swabbed and for major MOCs where rinsate samples will be collected during validation and tested using validated analytical methods.
- Establish dirty hold times, clean hold times, and maximum campaign length.
- Investigate and resolve any deviations during CV runs.
- Write and approve final report.

CLEANING PROCEDURES

Cleaning procedures shall have adequate detail to provide appropriate consistency of execution for the cleaning process. Adequate detail may include information to control critical cleaning process parameters such as time, cleaning action (including tools and technique), detergent type and concentration, and temperature as well as sequence of process steps.

SAMPLING METHODOLOGIES

Samples, either rinsate and/or direct surface swabbing, are obtained during the execution of the cleaning validation to determine the effectiveness of the cleaning procedure.

Direct sampling methods such as swabbing must be employed wherever practical and appropriate. Worst-case swabbing locations must be identified and the rationale specified in the protocols. Swab locations are identified based on the results from spray coverage studies, worst-case cleaning locations, hot spots, experience from the development studies, etc. Rinsate sampling may be employed if swabbing is difficult or routine disassembly for swabbing is not feasible.

PERFORMING AND DOCUMENTING CLEANING STUDIES

Cleaning Validation studies will be performed under approved protocols. The protocol is a written plan detailing the requirements for performing and evaluating cleaning validation studies.

Prior to initiating a cleaning validation study the following must be in place:

- Approved Cleaning Validation Procedure or Cycle
- Approved and validated analytical methods, including recovery studies
- CV limits established to assess the cleanliness of the equipment
- Approved Cleaning Validation protocol

Final reports are issued to document all CV activities including both acceptable and failed studies.

ACCEPTANCE CRITERIA

Systems are considered clean when the equipment is visually clean and potential soil loads and/or detergent levels have been reduced to or below established acceptance limits. It is important to identify these substances during execution of Stage 1 Cleaning Development activities and determine acceptable levels of carry-over (MAC). This establishes safe levels at which the residues would not be considered a contaminant in the next manufactured product. Different types of residues are described below:

- Residual product or degraded residual product remaining at the completion of manufacturing activities.
- Excipients or other inactive ingredients, unless endogenous or generally recognized as safe (GRAS), which may contaminate the subsequently manufactured product.
- Cleaning agents, such as detergents, are often used as aids to remove residues from equipment surfaces. Even though the detergent is soluble in the cleaning and rinsing solution, the reduction of these cleaning agent residues within acceptable limits must be evaluated in the cleaning validation study.
- Microbiological residues (bioburden and endotoxin) introduced during the manufacturing or storage process.

Visual Inspection

Visual inspection is performed on applicable equipment after a cleaning process has been completed. The purpose of the visual inspection is to ensure completeness of cleaning and associated documentation. Successful visual inspection is the first criteria to ensure cleanliness during validation and during routine production.

To view hard-to-reach areas, such as underneath the agitator blades, underneath the domes, etc., mirrors with extended poles may be used. Use of borescopes or other video devices are recommended to view hard-to-reach areas such as inside long pipes and other delivery systems.

Chemical Analysis

Active Ingredient Residues

The maximum amount of residue allowed in the next product after cleaning the previous product, in an analytical sample, or as an amount per surface area will be determined using validated analytical methods. Lowest of the values calculated in one of the following methods will be used as the residual limit for the potential residues.

- Levels based on scientific rationale not to exceed 0.001 of a minimum therapeutic daily dose of any product.
- The acceptable carryover calculated based on the Acceptable Daily Intake (ADI) of any pharmaceutical inactive component that has no dosage data.
- No more than 10 ppm of any product in the next product.

Cleaning Agent Residues

Any cleaning agents regulated by local agencies that have defined residual limits (e.g. EPA or OSHA), these limits shall be followed. For any cleaning agent that does not have a pre-defined limit, a limit must be determined. These limits may either be based upon a special report toxicology data, or matrix based. For matrix-based limits, a matrix rationale will be utilized and documented with the validation protocol specific to the cleaning process.

FAILURE OF A CLEANING VALIDATION

In the event a cleaning validation results in a failure, an investigation is conducted to determine if the failure is intrinsic or extrinsic.

INTRINSIC CAUSES

Intrinsic causes are those directly related to the cleaning procedure under study.

- Intrinsic causes require root cause identification and implementation of corrective and/or preventative actions prior to reinitiating cleaning validation for the product.
- Re-initiation of a CV study with changes in limits or procedure requires issuance of a new CV protocol. The protocol will include discussion of the failure and justification for the new study.
- A failure report will be issued for the failed validation study and will include discussion of the failure event, including an explanation of the issue resolution and an assessment of its impact on the CV study.

EXTRINSIC CAUSES

Extrinsic causes are external to the process undergoing validation, and resolution of the failure does not lead to changes in the cleaning process. An equipment failure during cleaning could be an example of an extrinsic cause.

- If an extrinsic cause is assigned for the failure, the run can be invalidated, and additional run(s) can be performed to complete the study requirements.

- Extrinsic causes require root cause identification and implementation of corrective and/or preventative actions, as appropriate, prior to issuance of additional batch(es) to complete the full complement of cleanings required for the study.
- The summary report includes a discussion of the failure event, including an explanation of the issue resolution and any impact on the CV study.

DEVIATIONS

Departures during cleaning study activities are handled as follows:

- Problem/Resolution Report in accordance with SOP
- The impact of the discrepancy is discussed in the CV final report.

REVALIDATION

In general, revalidation is required if a TACT parameter is reduced (e.g., less time, reduced cleaning agent concentration, lower temperature). Minor changes to cleaning recipes, procedures, or system configuration will be evaluated to determine if cleaning revalidation is required.

Significant changes to approved cleaning procedures may require revalidation based on risk assessment. Examples of significant changes are a change in the cleaning agent, significant reduction in the amount of cleaning agent, etc.

CONTINUOUS PROCESS VERIFICATION (STAGE 3)

Once the cleaning procedure for product/equipment has been successfully validated, a continuous process verification (CPV) plan is defined based upon risk assessment (GQP-07-24).

Periodic monitoring activities (CPV studies) will be performed to ensure the system remains in a validated state. These studies should include routine visual inspection, analytical testing, and microbial testing (for monitoring hold times) following completion of cleaning activities. CPV studies will be performed according to a pre-approved protocol for compliance with the validated state.

In the event of a failure, an investigation will be conducted to determine the root cause. Based on the findings of the investigation, re-validation or even re-development of the cleaning procedures may be required.

Index

Printed in the United States
by Baker & Taylor Publisher Services